DEJA REVIEW™
Biochemistry

NOTICE

Medicine is an ever-changing science. As new research and clinical experience broaden our knowledge, changes in treatment and drug therapy are required. The authors and the publisher of this work have checked with sources believed to be reliable in their efforts to provide information that is complete and generally in accord with the standards accepted at the time of publication. However, in view of the possibility of human error or changes in medical sciences, neither the authors nor the publisher nor any other party who has been involved in the preparation or publication of this work warrants that the information contained herein is in every respect accurate or complete, and they disclaim all responsibility for any errors or omissions or for the results obtained from use of the information contained in this work. Readers are encouraged to confirm the information contained herein with other sources. For example and in particular, readers are advised to check the product information sheet included in the package of each drug they plan to administer to be certain that the information contained in this work is accurate and that changes have not been made in the recommended dose or in the contraindications for administration. This recommendation is of particular importance in connection with new or infrequently used drugs.

DEJA REVIEW™

Biochemistry

Second Edition

Saad M. Manzoul, MD

Resident Physician
Diagnostic Radiology Program
George Washington University Hospital
Washington, DC

Hussan Mohammed

University of Maryland School of Medicine
Baltimore, Maryland
Class of 2012

 **Medical**

New York   Chicago   San Francisco   Lisbon   London   Madrid   Mexico City
Milan   New Delhi   San Juan   Seoul   Singapore   Sydney   Toronto

The *McGraw·Hill* Companies

Deja Review™: Biochemistry, Second Edition

1 2 3 4 5 6 7 8 9   DOC/DOC   14 13 12 11 10

ISBN 978-0-07-162717-7
MHID 0-07-162717-0

This book was set in Palatino by Glyph International.
The editors were Kirsten Funk and Christie Naglieri.
The production supervisor was Catherine Saggesse.
Production management was provided by Preeti Longia Sinha of Glyph International.

This book is printed on acid-free paper.

Library of Congress Cataloging-in-Publication Data

Manzoul, Saad M.
    Deja review. Biochemistry / Saad M. Manzoul.—2nd ed.
        p. ; cm.—(Deja review)
    Includes bibliographical references and index.
    ISBN-13: 978-0-07-162717-7 (pbk. : alk. paper)
    ISBN-10: 0-07-162717-0 (pbk. : alk. paper)   1. Biochemistry—Examinations, questions, etc.   I. Title.   II. Title: Biochemistry.   III. Series: Deja review.
    [DNLM: 1. Biochemical Phenomena—Examination Questions.   QU 18.2 M296d   2010]
    QP518.3.M34   2010
    572.0076—dc22
                                                            2009039025

McGraw-Hill books are available at special quantity discounts to use as premiums and sales promotions, or for use in corporate training programs. To contact a representative please e-mail us at bulksales@mcgraw-hill.com.

To Mahmoud A. Manzoul
—Saad M. Manzoul

To my parents, to my big brother Yasar, and
to my wife Muna
—Hussan S. Mohammed

Contents

Faculty Reviewers

Robert Haynie, MD, PhD
Associate Dean for Student Affairs
Case Western Reserve University
School of Medicine
Cleveland, Ohio

Michael W. King, PhD
Professor of Biochemistry and Molecular
  Biology
Indiana University School of Medicine
Center for Regenerative Biology and
  Medicine
Terre Haute, Indiana

Student Reviewers

Betty Chung
UMDNJ
School of Osteopathic Medicine
Class of 2011

Adam Darnobid
SUNY Upstate Medical University
Class of 2009

Lee Donner
SUNY Downstate College of Medicine
Class of 2009

Contributing Author

Ekow Mills-Robertson
University of Maryland School of Medicine
Baltimore, Maryland Class of 2011

Contributors to First Edition

Bruk Endale
Case Western Reserve University
School of Medicine
Cleveland, Ohio
Class of 2007

Delwin S. Merchant
Case Western Reserve University
School of Medicine
Cleveland, Ohio
Class of 2008

Anand Satyapriya
Case Western Reserve University
School of Medicine
Cleveland, Ohio
Class of 2007

Preface

The *Deja Review* series is a unique resource that has been designed to allow you to review the essential facts and determine your level of knowledge on the subjects tested on Step 1 of the United States Medical Licensing Examination (USMLE). This second edition of *Deja Review: Biochemistry* was designed for the student as a compact yet high-yield review of the major biochemical concepts necessary to master the subject in how it relates to medicine.

ORGANIZATION

All concepts are presented in a question and answer "flashcard" format that covers key facts on commonly tested topics in biochemistry. This book not only manages to cover the basic topics but also integrates pathophysiology and pharmacologic correlates with newly introduced material. We believe that, along with the easy-to-understand figures, this book will serve as a great review and an important tool in assessing one's level of knowledge and achievement in biochemistry.

The question and answer format has several important advantages:

- It provides a rapid, straightforward way for you to assess your strengths and weaknesses.
- It serves as a quick, last-minute review of high-yield facts
- It allows you to efficiently review and commit to memory a large body of information.

At the end of each chapter, you will find clinical vignettes that expose you to the prototypic presentation of diseases classically tested on the USMLE Step 1. These board-style questions put the basic science into a clinical context, allowing you to apply the facts you have just reviewed in a clinical scenario and "Make the Diagnosis."

HOW TO USE THIS BOOK

This text was assembled with the intent to represent the core topics tested on course examinations and USMLE Step 1. Remember, this text is not intended to replace comprehensive textbooks, course packs, or lectures. It is simply intended to serve as a supplement to your studies during your Biochemistry course work and Step 1 preparation. You may use the book to quiz yourself or classmates on topics covered in recent lectures and clinical case discussions.

However you choose to study, we hope you find this resource helpful throughout your preparation for course examinations and the USMLE Step 1.

Saad M. Manzoul
Hussan Mohammed

Acknowledgments

The authors would like to thank each of the contributors, Dr. Robert Haynie, and Dr. Michael King for their time and effort in organizing and reviewing the information contained in this book. For their knowledge and expertise throughout the entire process, we would also like to thank Kirsten Funk, Marsha Loeb, and the entire McGraw-Hill staff.

I would like to thank the entire Manzoul family, Nagwa Taha, the Mohammed family, the Elhagmusa family, the Elbuluk family, and my friends for their unwavering support and encouragement.

Saad M. Manzoul

My sincere thanks to Saad for giving me the opportunity to work with him on this book. I would also like to thank my family, the Abdalla family, and all of my friends for their steadfast encouragement.

Hussan S. Mohammed

CHAPTER 1

Proteins

OVERVIEW

How many different amino acids make up mammalian proteins?

20

How do amino acids connect to make proteins?

Via peptide bonds (amino group bound to carboxyl group)

What is the basic structure of an amino acid?

Amino group + carboxylic acid group (Fig. 1.1)

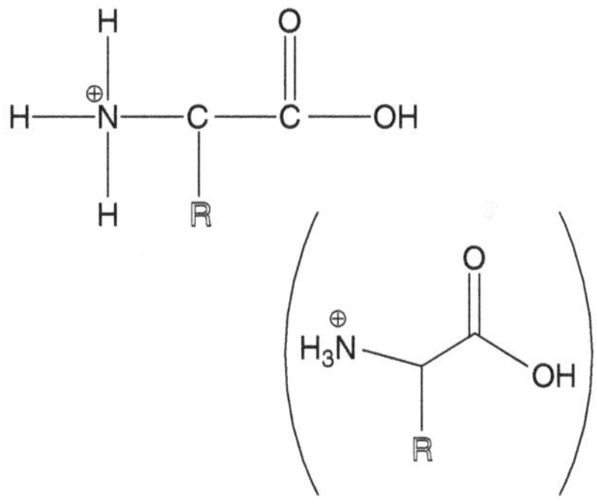

Figure 1.1

What are four classes of amino acid side chains?

1. Amino acids with nonpolar side chains
2. Amino acids with uncharged polar side chains
3. Amino acids with acidic side chains
4. Amino acids with basic side chains

Name the amino acids with nonpolar side chains:

Glycine, alanine, valine, leucine, isoleucine, phenylalanine, tryptophan, methionine, proline

1

Name the amino acids with uncharged polar side chains:	Serine, threonine, tyrosine, asparagine, cysteine, glutamine
Name the amino acids with acidic side chains:	Aspartic acid, glutamic acid
Name the amino acids with basic side chains:	Histidine, lysine, arginine
What are the essential amino acids?	**MILK WiTH** Veggies and Fruits is essential: **M**—methionine, **I**—isoleucine, **L**—leucine, **K**—lysine, **W**—tryptophan, **T**—threonine, **H**—histidine, **V**—valine, **F**—phenylalanine

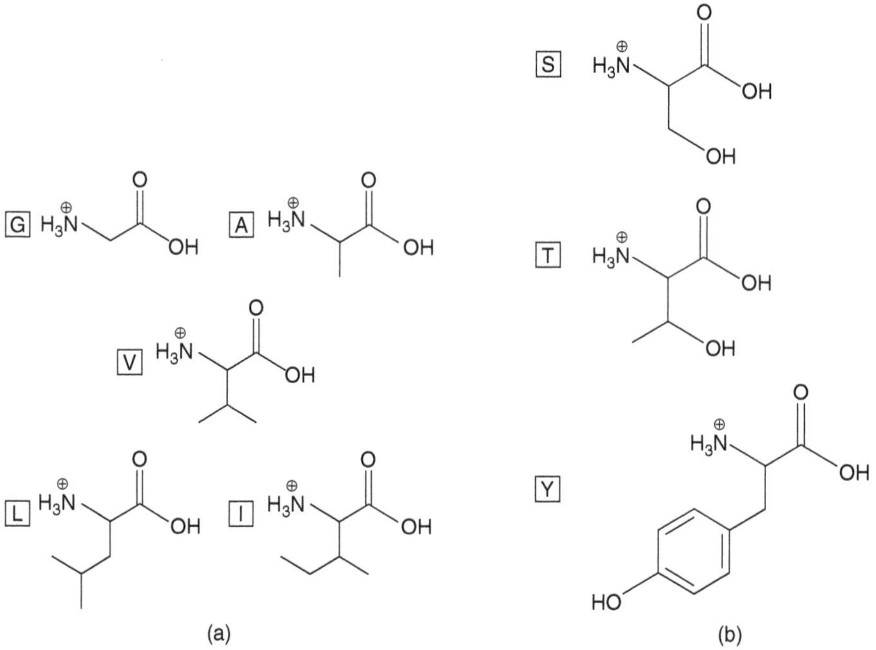

Figure 1.2   Amino acids. (a) Alipathic side chains (G—Glycine, A—Alanine, V—Valine, L—Leucine, I—Isoleucine); (b) Hydroxylic group side chains (S—Serine, T—Threonine, Y—Tyrosine); (c) Sulfur-containing side chains (C—Cysteine, M—Methionine); (d) Acidic side chains (D—Aspartic acid, N—Asparagine, E—Glutamic acid, Q—Glutamine); (e) Basic side chains (R—Arginine, K—Lysine, H—Histidine); (f) Aromatic ring-containing side chains (H—Histidine, F—Phenylalanine, Y—Tyrosine, W—Tryptophan); (g) (P—Proline).

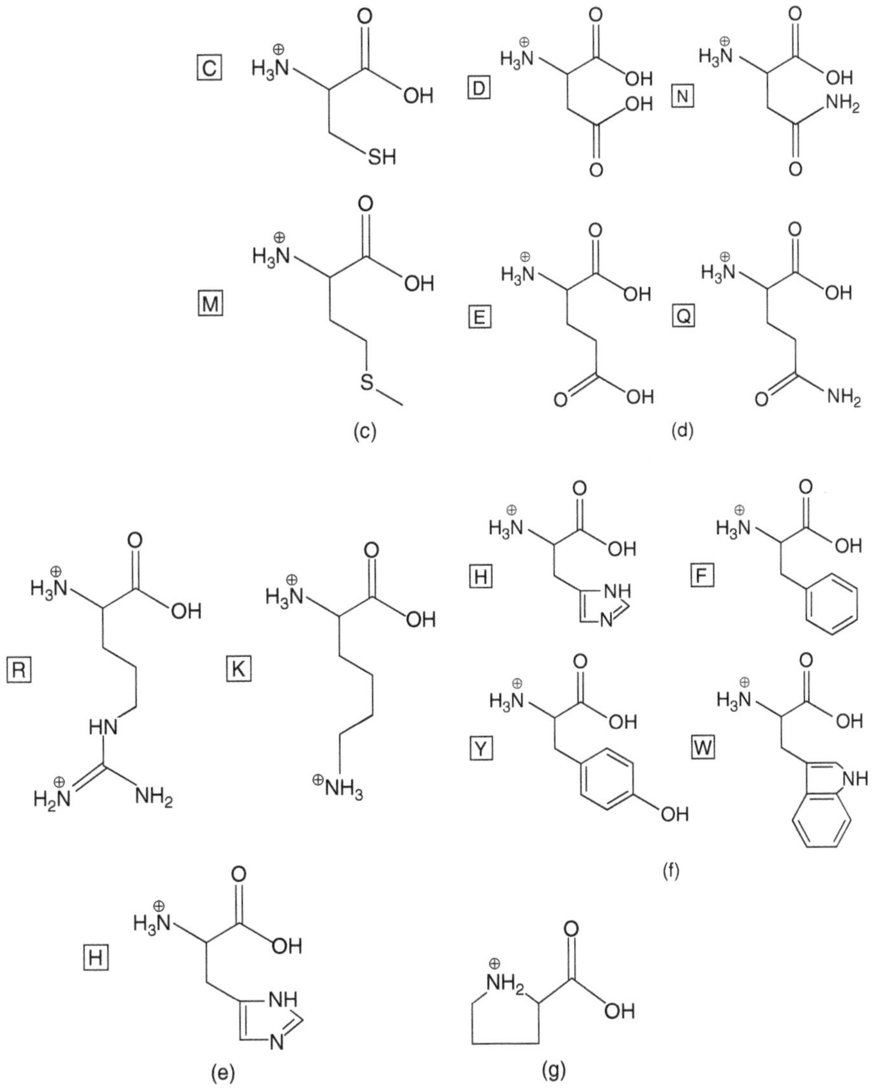

Figure 1.2 (*Continued*)

What is the Henderson-Hasselbalch equation?

pH = pKα + log [A⁻]/[HA] (where A⁻ is the acid form and HA is the base form)

How does this equation pertain to amino acids?

At pH values greater than the pKα, the carboxyl group of an amino acid loses its proton (H⁺), COO⁻; at pH values less then the pKα, an amino group gains a proton (H⁺).

What is physiologic pH?	7.35–7.45 (blood pH = 7.4; the kidneys and lungs work together to maintain it)
What is considered acidotic pH?	Less than 7.35
What is considered alkalotic pH?	Greater than 7.45
For amino acids with uncharged side chains, what is their overall charge?	They are neutral at physiologic pH due to the amino group being positive (NH_3^+) and the carboxyl group being negative (COO^-).
At an acidic pH, what is the charge of an amino acid with an uncharged side chain?	Positive due to the amino group gaining a proton (NH_3^+)
At an alkalotic pH, what is the charge of an amino acid with an uncharged side chain?	Negative due to the carboxyl group losing a proton (COO^-)
What are three characteristics of the peptide bond?	1. Lack of rotation around the bond 2. Planar 3. Uncharged but polar
What are the four levels of protein structure?	1. Primary 2. Secondary 3. Tertiary 4. Quaternary
What does primary structure mean?	The linear sequence of amino acids in a protein
What is the secondary structure of a protein?	The regular arrangements of amino acids; the 3-D structure of one or more stretches of amino acids
What are the two most common types of secondary structure?	α-Helix and β-sheet
Name two proteins in the body that are primarily α-helical:	1. Hemoglobin (80% α-helical) 2. Keratin (nearly entirely α-helical)
Name two properties of an α-helix:	1. It is stabilized by hydrogen bonds between the peptide-bond carbonyl oxygens and amide hydrogens that are part of the polypeptide backbone. 2. Each turn of the α-helix contains approximately four amino acids.

Which amino acids disrupt an α-helix?	Proline (inserts a kink in the chain), large numbers of charged amino acids (glutamate, aspartate, histidine, lysine arginine), amino acids with bulky side chains (tryptophan), or amino acids that branch at the β-carbon (valine, isoleucine)
What is the difference between an α-helix and a β-sheet?	An α-helix contains one peptide chain; a β-sheet contains two or more.
What are the forms a β-sheet can take on?	They can either be parallel or antiparallel β-sheets.
Which β-sheet form gives more overall stability?	Antiparallel arrangements give more stability than parallel ones.
Name a protein-like structure composed of primarily β-sheets:	Amyloid
In what disease processes is amyloid deposited within the body?	Alzheimer disease, multiple myeloma, Down syndrome, chronic hemodialysis
What does tertiary structure mean?	The folding of domains, and the final arrangement of domains in the polypeptide
What determines the tertiary structure of a protein?	The primary structure (the sequence of amino acids) and the interactions of the primary structure
Name the four types of interactions that cooperate in stabilizing the tertiary tructure of a protein:	1. van der Waals forces 2. Hydrophobic interactions 3. Hydrogen bonds 4. Ionic interactions
What is the quaternary structure of a protein?	The arrangement of the many subunits in a protein (if two subunits—dimeric, three—trimeric, and so on)
What is the quaternary structure of hemoglobin?	Hemoglobin is a heterotetramer; four globular protein subunits
What is protein denaturation?	The unfolding and disorganization of a protein's structure; secondary and tertiary structures are disrupted but the peptide bonds between amino acids are left intact.

| What can cause the denaturation of a protein? | Heat, organic solvents, mechanical mixing, strong acids or bases, detergents, and ions of heavy metals such as lead and mercury |

HEMEPROTEINS

| What are the two most abundant hemeproteins in humans? | Hemoglobin and myoglobin |

| What is the difference between hemoglobin and myoglobin? | Hemoglobin is made up of four subunits, whereas myoglobin is made up of one subunit. Hemoglobin is found in red blood cells (RBCs) and myoglobin is found in muscle. Hemoglobin also has less affinity for oxygen than myoglobin. |

| What is heme? | Heme is a complex of protoporphyrin IX and ferrous ion (Fe^{2+}) (becomes oxidized to Fe^{3+} when oxygen is bound) (Fig. 1.3) |

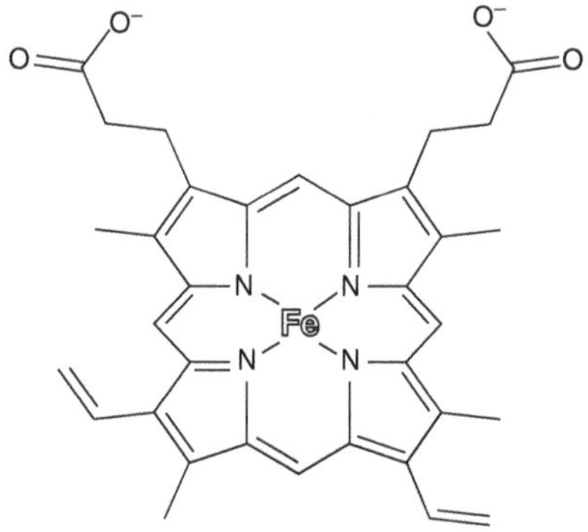

Figure 1.3

| What enzyme catalyzes the rate-limiting step of heme synthesis? | Aminolevulinate (ALA) synthase, found in the liver and bone marrow |

| How does lead cause microcytic, hypochromic anemia, and porphyria? | Lead inhibits aminolevulinate synthase and ferrochelatase, thus preventing the incorporation of iron into heme. |

What are the different types of β-thalassemia?

One of the two genes is defective = β-thalassemia minor

Both of the genes are defective = β-thalassemia major

In what geographical regions is α-thalassemia most prevalent?

Asia and Africa

In which geographical region is β-thalassemia most prevalent?

Mediterranean

What type of hemoglobin is increased in β-thalassemia?

Hgb F

What is the treatment for β-thalassemia major?

Blood transfusion

What is a complication of this treatment?

Cardiac failure due to secondary hemochromatosis

Characterize the anemia associated with thalassemia:

Microcytic hypochromic anemia

What sign on x-ray is indicative of thalassemia?

"Crew cut" on skull x-ray due to marrow expansion (as in sickle cell anemia)

CONNECTIVE TISSUE PROTEINS

Name the connective tissue proteins found throughout the body:

Collagen, elastin, and keratin

Where are these proteins located in the body?

Collagen and elastin are found in connective tissue, sclera, cornea, and blood vessel walls; keratin is found in skin and hair.

What is the most abundant protein in the body?

Collagen

What is the structure of collagen?

Three polypeptides (α-chains) wrapped around each other in a triple-helix formation

What is the sequence of amino acids in the triple helix of collagen?

Glycine-XY (X and Y may be proline, hydroxyproline, or hydroxylysine)

What vitamin is required for the synthesis of collagen?	Vitamin C, required for hydroxylation of proline residues in collagen (technically vitamin C is not required for the synthesis of collagen, but posttranslational modification)
Deficiency of vitamin C is referred to as what disease?	Scurvy
Name common physical findings associated with this disease:	Swollen gums, easy bruising, anemia, poor wound healing, weakness
How is collagen synthesized?	Specific prolyl and lysyl residues are hydroxylated via prolyl hydroxylase in the endoplasmic reticulum, forming procollagen. Procollagen is exocytosed into the extracellular space. Procollagen is then cleaved to form tropocollagen, which then aggregates into collagen fibrils.
What characteristic linking reinforces the structure of collagen?	Covalent cross-linking of lysine-hydroxylysine residues between tropocollagen molecules
Where is collagen type I located in the body?	Bone, tendon, skin, dentin, fascia, cornea, late wound repair; 90% of collagen in the body is type I collagen (remember b**ONE**)
Where is collagen type II located in the body?	Cartilage, vitreous body, nucleus pulposus (remember car**TWO**lage)
Where is collagen type III located in the body?	Reticulin fibers located in skin, blood vessels, uterus, fetal tissue, granulation tissue (remember re**THREE**culin)
Where is collagen type IV located in the body?	Basement membrane or basal lamina (remember under the F(l)**OUR**)
Where is collagen type X located in the body?	Epiphyseal plate (remember cartilage calcifica**TEN**)
Name three diseases which result from dysfunctional collagen synthesis:	1. Scurvy 2. Ehlers-Danlos syndrome 3. Osteogenesis imperfecta
Name several signs of Ehlers-Danlos syndrome:	Hyperextensible skin, tendency to bleed (easy bruising), hypermobile joints

How many different types of Ehlers-Danlos syndromes are there?	About 10 types (including those with autosomal recessive, autosomal dominant, and X-linked recessive inheritance patterns)
What significant vascular pathology is associated with Ehlers-Danlos syndrome?	Intracranial berry aneurysms
What is the most common mode of inheritance for osteogenesis imperfecta?	Autosomal dominant
What are some signs and symptoms of osteogenesis imperfecta?	Multiple fractures with minimal trauma, blue sclerae, hearing loss (abnormal middle ear bone formation), and dental problems (lack of dentition)
Osteogenesis imperfecta may often be confused with what suspicion on the part of the physician?	Child abuse
Which type of osteogenesis imperfecta is fatal?	Type II (fatal in utero or in the neonatal period)
What autoimmune disease is due to antibodies against the basement membrane (collagen type IV)?	Goodpasture syndrome
Describe the clinical manifestations of the above disease:	Goodpasture primarily affects the lungs and kidneys; symptoms may include coughing up blood, bloody or dark-colored urine, burning sensation during urination, difficulty breathing, nausea, pale skin, and fatigue.
What genetic disorder prevents the proper production of the basement membrane (collagen type IV)?	Alport syndrome
Describe the clinical manifestations of the above disease:	Alport syndrome primarily affects the kidneys, ears, and eyes; symptoms may include progressive hereditary nephritis, bloody urine, bilateral hearing loss, and ocular lesions.
Which serum protein inhibits elastin degradation?	α-1-Antitrypsin inhibits neutrophil elastase (a protease that acts in the extracellular space to degrade the elastin of alveolar walls and other structural proteins).

Where is α-1-antitrypsin produced?	Primarily in the liver, by monocytes, and macrophages
What diseases are associated with α-1-antitrypsin deficiency?	Emphysema (barrel-chested individuals suffering from air trapping due to increased compliance but decreased elasticity) and liver damage (cirrhosis, cholestasis)

ENZYMES

What are enzymes?	Protein catalysts that increase the rate of chemical reactions without being altered in the process
Name six properties of an enzyme:	1. Contains an active site 2. Catalytically efficient 3. Substrate-specific 4. Utilizes cofactors 5. Activity is potentially regulated 6. Located in specific areas of the cell or extracellular space
Does an enzyme change the chemical equilibrium of a reaction?	No, it allows the reaction to take place at a faster rate by decreasing the energy required to start a reaction.
Name the factors that can affect the rate of enzymatic catalysis of an enzyme:	Substrate concentration, pH, and temperature

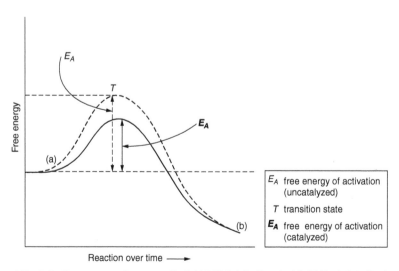

Figure 1.8   Activation energy and enzyme effect. (a) Initial state (Reactants); (b) Final state (Products).

What is the effect of temperature on the reaction velocity?

Increases with temperature until a peak velocity is reached and declines

What is the effect of substrate concentration on reaction velocity?

Increases with increasing substrate concentration until a maximal velocity is reached

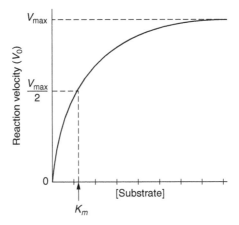

Figure 1.9   Substrate concentration versus reaction velocity.

What is V_{max}?

The maximum velocity at which an enzyme can catalyze a reaction (all conditions are optimal)

What is the effect of pH on enzymatic activity?

The catalytic process usually requires the enzyme (E) and substrate (S) to have certain chemical groups in an ionized or unionized state so they can react, that is, some enzymes may require an amino group to be protonated (NH_3^+). Thus, at a high (basic) pH, the amino group is deprotonated and the enzyme-substrate (ES) complex cannot form; furthermore, extremes of pH can denature the enzyme.

$$E + S \underset{k_{-1}}{\overset{k_1}{\rightleftharpoons}} ES \overset{k_2}{\longrightarrow} E + P$$

Figure 1.10   Enzymatic action.
(a) E—Enzyme; (b) S—Substrate;
(c) ES—Enzyme-substrate complex;
(d) k_1, k_{-1}, k_2—Rate constants.

What is the Michaelis-Menten equation?	$V_0 = V_{max}[S]/(K_m + [S])$ V_0 = initial reaction velocity V_{max} = maximal velocity K_m = Michaelis constant $[S]$ = substrate concentration
What does the Michaelis-Menten equation represent?	The relationship between initial reaction velocity and substrate concentration
What are the three assumptions of the Michaelis-Menten equation?	1. $[S]$ is much greater than $[E]$. 2. $[ES]$ does not change with time; that is, the rate of ES formation is equal to the rate of ES breakdown to either $E + S$ or $E + P$. 3. Only the initial velocities are assumed.
What is K_m?	K_m is equal to the substrate concentration at which the reaction velocity is equal to $1/2\ V_{max}$.
What does a small K_m imply?	High affinity of the enzyme for the substrate
What does a large K_m imply?	Low affinity of the enzyme for the substrate, meaning a high concentration of substrate is needed to half-saturate the enzyme
Does hexokinase have a large or small K_m?	Low K_m, thus high affinity for glucose; it is saturated at normal blood glucose concentrations; hexokinase is found in most tissues
Does glucokinase have a large or small K_m?	High K_m, thus low affinity for glucose; it is saturated only at high blood glucose concentrations; glucokinase is present in only the liver and pancreas
What is meant by the term *zero-order reaction*?	The velocity of the reaction is constant and independent of substrate concentration.
What enzyme showcases zero-order reaction kinetics?	Alcohol dehydrogenase in ethanol metabolism; you cannot speed up the reaction by adding more enzyme

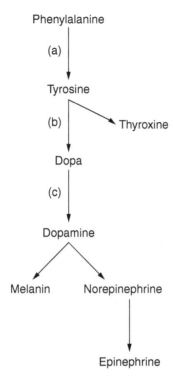

Figure 1.12   Phenylalanine metabolism. (a) Phenylalanine hydroxylase; (b) Tyrosine hydroxylase; (c) Dopa decarboxylase.

What substances are found in increased amounts in the urine of those with PKU?	Phenylketones (i.e., phenyllactate, phenylpyruvate, phenylacetate)
When is PKU screened?	At birth
PKU is inherited in what pattern?	Autosomal recessive
What is the treatment for PKU?	Decreased phenylalanine, increased tyrosine in the diet
What is maternal PKU?	This occurs when a mother's uncontrolled hyperphenylalaninemia leads to mental retardation and birth defects in the fetus as a result of phenylalanine crossing the placenta and interfering with organogenesis during weeks 3–8 of gestation. It may occur even if the fetus is not deficient in phenylalanine hydroxylase or tetrahydrobiopterin cofactor.

Describe the pathophysiology behind maple syrup urine disease:	Degradation of branched amino acids (i.e., isoleucine, valine, leucine) is blocked, thus causing increased α-ketoacids in the blood.
What is the deficient enzyme in maple syrup urine disease?	Branched-chain ketoacid dehydrogenase (cofactor thiamine)
What are the symptoms of maple syrup urine disease?	Mental retardation, edema, white matter demyelination, hyperglycemia, and death
What pathophysiological processes can result in albinism?	Deficiency of tyrosinase Defective tyrosine transporters (which decreases the amount of melanin) Lack of migration of neural crest cells
What is the role of tyrosinase?	Aids in the conversion of tyrosine to melanin
What is the role of melanin?	Gives pigment to the hair, skin, and iris
Which important pathology is associated with a decrease in melanin?	Increased susceptibility to sunburns with an increased risk of developing squamous cell carcinoma
Describe the pathophysiology behind homocystinuria:	Defective cystathionine synthase and/or methionine synthase resulting in excess homocystine in the urine
Describe the two forms of homocystinuria:	1. Deficiency of either of the aforementioned enzymes 2. Decreased affinity of either enzyme for pyridoxal phosphate (vitamin B_6)
What is the treatment for the first form?	Increased cysteine (because cysteine becomes an essential amino acid) and decreased methionine (because there is an accumulation of methionine and its metabolites in the blood) in the diet
What is the treatment for the second form?	Increased vitamin B_6 in the diet

What are the symptoms of homocystinuria?	Marfan-like body habitus (i.e., tall stature with long extremities, hyperextensive joints, long and tapering fingers and toes, lens dislocations), mental retardation, osteoporosis, thrombotic episodes (as a result of homocysteine damaging the vascular endothelium)
What physiological defect is found in cystinuria?	Defective renal tubular amino acid transporter for cystine, ornithine, lysine, and arginine (i.e., **COLA**)
What major symptom is associated with cystinuria?	Cystine kidney stones (radiolucent stones seen via imaging, yellow-brown hexagonal crystals seen macroscopically)
What is the treatment for cystinuria?	Acetazolamide (to alkalinize the urine and allow further excretion of cystine)
What is the incidence of cystinuria?	About 1:7000; important because it is the most common genetic error of amino acid transport
What is the deficient enzyme in alkaptonuria?	Homogentisic acid oxidase
The above enzyme is utilized in the degradation of which amino acid?	Tyrosine

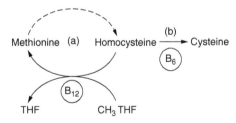

Figure 1.13 Homocysteine metabolism. (a) Methionine synthase; (b) Cystathionine synthase.

What are the symptoms of alkaptonuria?	Generally no symptoms other than dark urine and connective tissue (onchronosis), although rarely some have severe arthralgias

What causes the urine and connective tissue to be dark?	Homogentisic acid accumulates, forming polymers (alkapton bodies), which causes the urine and connective tissue to darken
What is the prevalence of alkaptonuria?	1:250,000
What is the deficient enzyme in cystathioninuria?	Cystathionase
What reaction does cystathionase catalyze?	Formation of cysteine and α-ketobutyrate from cystathionine
What are the symptoms of cystathioninuria?	No clinical symptoms
What is the deficient enzyme in histidinemia?	Histidase
What reaction does this enzyme catalyze?	Histidine to urocanate
What are the symptoms of histidinemia?	Often asymptomatic, but mental retardation may be present
Describe the pathophysiology behind histidinemia:	Symptoms occur due to increased levels of histidine in the blood and urine
What inheritance pattern is seen in histidinemia?	Autosomal recessive
What is the prevalence of histidinemia?	1:10,000

CLINICAL VIGNETTES—MAKE THE DIAGNOSIS

A 2-year-old is brought to the pediatrician for a well-child visit. The child is noted to have white hair, including eyelashes and eyebrows. Her skin is very pale throughout and eye examination reveals poor development of the macula with irises and pupils. The child's father has many of the same characteristics.

This child is suffering from what disorder?	Albinism (absence of tyrosinase)
The absent enzyme catalyzes what reaction?	Tyrosinase catalyzes the conversion of tyrosine to dihydroxyphenylalanine and melanin
The child may be at risk for developing what skin disease?	Squamous cell cancer

CHAPTER 2

Metabolism

OVERVIEW

What are the three major classes of carbohydrates?	1. Monosaccharides 2. Disaccharides 3. Polysaccharides
What is the simplest of these carbohydrates?	Monosaccharides
What is the major fuel source of the brain?	Glucose
What are the major metabolic pathway(s) of the brain?	Glycolysis and amino acid metabolism
What cells do not contain mitochondria and thus rely only on glycolysis for energy production?	Erythrocytes
What type of tissue stores, synthesizes, and mobilizes triglycerides?	Adipose tissue
Which metabolic pathway does fast-twitch muscle use for fuel?	Glycolysis
Which metabolic pathway does slow-twitch muscle use for fuel?	Tricarboxylic acid (TCA) cycle and β-oxidation (i.e., aerobic pathways)

GLYCOLYSIS

Name the family of glucose carrier proteins that transport glucose into the cell:	The GLUT proteins
What glucose transporter is used by the liver?	GLUT2

What glucose transporter is used by muscle?	GLUT4
Which one of the above transporters is sensitive to insulin?	GLUT4
What is the mechanism of action of insulin on this transporter?	Facilitates movement of the transporter to cell membrane
What glucose transporter is located on the brush-border membrane of both intestinal and kidney cells?	GLUT2
The above enzyme is coupled to the transport of what ion to provide energy for glucose transport?	Na^+
In most tissues, glucose is trapped in the cell by phosphorylation by what enzyme?	Hexokinase
What inhibits the above enzyme?	Feedback inhibition by its product glucose-6-phosphate
In the liver, glucose is phosphorylated by what enzyme?	Glucokinase
What is the major distinction between hexokinase and glucokinase?	Glucokinase differs from hexokinase in that it requires a much larger glucose concentration (K_m) to achieve half saturation.
Does glucokinase or hexokinase prevent hyperglycemia following a carbohydrate-rich meal?	Glucokinase functions to prevent hyperglycemia following a carbohydrate-rich meal.
Which two organs express glucokinase?	1. Liver 2. Pancreas
Describe the kinetics of glucokinase:	It has a high K_m and high V_{max} and is not subject to feedback inhibition by glucose-6-phosphate.
Describe the kinetics of hexokinase:	It has a low K_m and low V_{max} and is subject to feedback inhibition by glucose-6-phosphate.

What is the effect of insulin on this glucokinase?

Insulin induces synthesis of the glucokinase.

Name two functions of glycolysis:

1. Degrading glucose to generate adenosine triphosphate (ATP)
2. Providing building blocks for synthetic reactions (such as the formation of long-chain fatty acids)

How much ATP is consumed per mole of glucose that undergoes glycolysis?

2 moles are consumed.

How much ATP is generated per mole of glucose that undergoes glycolysis?

4 moles

What is the net generation of ATP per mole of glucose that undergoes glycolysis?

2 moles

What is the major regulatory enzyme in glycolysis?

Phosphofructokinase-I (PFK-I)

Name the three enzymes of glycolysis that catalyze virtually irreversible reactions:

1. Hexokinase
2. PFK-I
3. Pyruvate kinase

What reaction does PFK-I catalyze?

Fructose-6-phosphate $\rightarrow$ fructose-1,6-bisphosphate (coupled to the hydrolysis of ATP)

Name a positive allosteric regulator of this enzyme:

Adenosine monophosphate (AMP), fructose-2,6-bisphosphate

Name an allosteric inhibitor of this enzyme:

ATP, citrate

What reaction does PFK-II catalyze?

Fructose-6-phosphate $\rightarrow$ fructose-2,6-bisphosphate

In what organ is PFK-II not regulated by phosphorylation?

Muscle

Is activity of PFK-II a sign of the fed or fasting state?	Fed state
Which two glycolytic intermediates liberate enough energy for driving ATP synthesis?	1. 1,3-Bisphosphoglycerate 2. Phosphoenolpyruvate (PEP)
What are the two ATP-producing enzymes of glycolysis?	1. 3-Phosphoglycerate kinase 2. Pyruvate kinase (Tip: remember "kinase")
Pyruvate kinase catalyzes what reaction?	PEP → pyruvate
What covalent modification inhibits pyruvate kinase?	Phosphorylation
What enzyme carries out the above allosteric inhibition?	Protein kinase A
Name the allosteric inhibitors of pyruvate kinase:	ATP, acetyl coenzyme (CoA); alanine in liver only
Name the allosteric activator of pyruvate kinase:	Fructose-1,6-bisphosphate
What are the signs of pyruvate kinase deficiency?	Anemia, reticulocytosis with macrovalocytosis, increased 2,3-bisphosphoglycerate (BPG) (Tip: remember red blood cells [RBCs] metabolize glucose anaerobically and thus depend solely on glycolysis)
This disorder is inherited in what pattern?	Autosomal recessive
What is the most effective treatment for this disorder?	Exchange transfusions
Which enzyme produces nicotinamide adenine dinucleotide (NADH) in glycolysis?	Glyceraldehyde-3-phosphate dehydrogenase
How much NADH is produced per mole of glucose oxidized to pyruvate?	2 moles

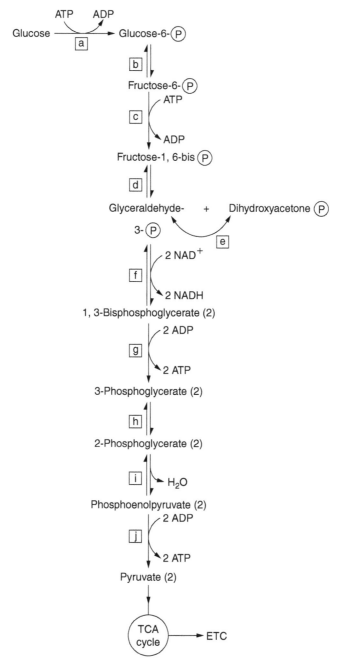

Figure 2.1   Glycolysis pathway. (a) Hexokinase/glucokinase; (b) Phosphoglucose mutase; (c) Phosphofructokinase I; (d) Aldolase; (e) Triose phosphate isomerase; (f) Glyceraldehyde-3-phosphate dehydrogenase; (g) Phosphoglycerate kinase; (h) Phosphoglycerate mutase; (i) Enolase; (j) Pyruvate kinase.

Since erythrocytes do not contain mitochondria, what is the NADH produced in glycolysis used for?	To reduce pyruvate to lactate
How is the reducing power of NADH transferred to the mitochondria?	Via the glycerol-3-phosphate shuttle or malate aspartate shuttle
What are the possible fates of pyruvate produced in the cell?	It can be converted to lactate, alanine, acetyl CoA, oxaloacetate, or glucose.
How many moles of ATP are required to generate glucose from pyruvate?	6 moles
Under anaerobic conditions, pyruvate is converted to what molecule?	Lactate (anaerobic conditions result in less ATP production than aerobic conditions)
What enzyme catalyzes the aforementioned reaction?	Lactate dehydrogenase
What are the three enzymes of the pyruvate dehydrogenase (PDH) complex?	1. Pyruvate decarboxylase 2. Dihydrolipoyl transacetylase 3. dihydrolipoyl dehydrogenase
What reaction does the PDH complex catalyze?	Pyruvate + NAD^+ + CoA $\rightarrow$ acetyl CoA CO_2 + NADH
What coenzymes are required by this enzyme?	Thiamine pyrophosphate (vitamin B_1), coenzyme A (vitamin B_5), NAD^+ (vitamin B_3), flavin adenine dinucleotide (FAD) (vitamin B_2), and lipoic acid
The PDH complex is similar to what other enzyme?	α-Ketoglutarate dehydrogenase complex
Is the PDH complex active in the phosphorylated or nonphosphorylated state?	Nonphosphorylated state
What enzyme phosphorylates the PDH complex?	PDH kinase
What molecules activate PDH kinase (thus inhibiting the PDH complex)?	Acetyl CoA, NADH

How does this molecule stimulate gluconeogenesis?

Excess acetyl CoA activates pyruvate carboxylase, increasing gluconeogenesis.

Which by-product of exercising or ischemic muscle is used for gluconeogenesis?

Lactate

Which process describes the movement of gluconeogenic substrates between the muscle and the liver?

Cori cycle (Fig. 2.3)

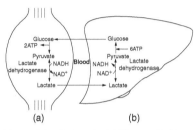

Figure 2.3   Cori cycle. (a) Muscle; (b) Liver.

OXIDATIVE PHOSPHORYLATION

Where in the eukaryotic cell is oxidative phosphorylation (i.e., the electron transport chain, ETC) carried out?

Across the mitochondrial inner membrane

What reagents are involved in oxidative phosphorylation?

Either NADH or FADH$_2$, which come from the TCA cycle

What do the aforementioned reagents contribute to the ETC?

Electrons

What are the products of oxidative phosphorylation?

H$_2$O; electrons are passed down the intermediates of the ETC and eventually reduce oxygen to produce H$_2$O

What is the purpose of oxidative phosphorylation?

To create a proton gradient by pushing H$^+$ into the mitochondrial intermembrane space, which then flows back into the mitochondrial matrix providing the energy to form ATP from ADP and P$_i$

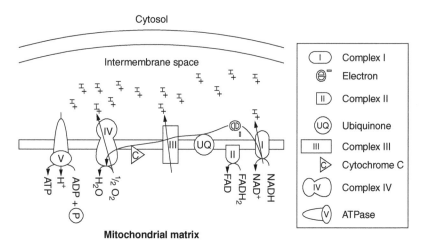

Figure 2.4   The electron transport chain.

How is the ETC coupled with oxidative phosphorylation?	The flow of ions into the mitochondrial matrix is considered to be oxidative phosphorylation, and the gradient is created by the ETC. Thus, these two processes are said to be coupled.
How many protein complexes are involved in the ETC?	Four complexes (Complexes I-IV) and two mobile electron carriers
Name the two mobile electron carriers:	1. Coenzyme Q (ubiquinone) 2. Cytochrome c
What is the name of Complex I?	NADH dehydrogenase
Electrons are transferred to Complex I of the ETC from what molecule?	$NADH_2$
What is the name of Complex II?	Succinate dehydrogenase
Where in the ETC does $FADH_2$ transfer its electrons?	Complex II (this allows only two ATPs to be made from $FADH_2$ since Complex I is bypassed)
Describe the flow of electrons from NADH through the ETC complexes:	NADH → Complex I → ubiquinone (coenzyme Q) → Complex III → cytochrome c → Complex IV → O_2
Describe the flow of electrons from $FADH_2$ through the ETC complexes:	$FADH_2$ → Complex II → ubiquinone (coenzyme Q) → Complex III → cytochrome c → Complex IV → O_2

Which complex transfers its electrons to O_2?	Complex IV
Which molecule transfers electrons from Complex I to Complex II?	Ubiquinone (also called coenzyme Q)
Which molecule transfers electrons from Complex III to Complex IV?	Cytochrome c
What is the only electron carrier that is not linked to a protein?	Coenzyme Q
As electrons pass through the complexes and the two mobile carriers, which complexes push H^+ ions into the intermembrane space?	Complexes I, III, and IV (Complex II does not push protons into the intermembrane space)
How much ATP is produced ultimately from NADH traveling down the ETC?	~3.5 ATP
How much ATP is produced ultimately from $FADH_2$ traveling down the ETC?	~2 ATP
What enzyme synthesizes ATP in the inner mitochondrial membrane?	The F_1F_0 ATPase
The energy for synthesis of ATP is provided by what?	Movement of protons down their concentration gradient
Which domain contains the proton-conducting channel?	The F_0 domain
Which domain is the site of ATP synthesis?	The F_1 domain
Approximately how many H^+ ions does one turn of the F_1F_0 ATPase require?	12–14 H^+ ions
What determines the rate of oxidative phosphorylation?	The availability of ADP
What is this type of regulation called?	Respiratory control, due to the fact that more ADP is created by increasing the metabolic rate (or *respiratory rate*), thus using up ATP and creating ADP
Name the inhibitors of Complex I:	Amobarbital (a barbiturate), rotenone (an insecticide), and piericidin A (an antibiotic)
Name an inhibitor of Complex II:	Antimycin A (an antibiotic)

Which oxygen analog inhibits the ETC?	Cyanide (CN)
Where does cyanide act in the ETC?	Complex IV
Name some other inhibitors of this complex in the ETC:	Hydrogen sulfide (H_2S) and carbon monoxide (CO)
What inhibits ATP synthesis by directly inhibiting the F_1F_0 ATPase?	Oligomycin
Describe its mechanism of action:	Oligomycin uncouples the ETC and oxidative phosphorylation by binding to the F_1F_0 ATPase; thus, oligomycin prevents the reentry of H^+ ions into the mitochondrial matrix and does not allow ATP to be formed.
What agent inhibits oxidative phosphorylation by disrupting the proton gradient across the inner mitochondrial membrane?	2,4-Dinitrophenol (DNP)
How does this molecule affect the proton gradient?	Makes the inner mitochondrial membrane permeable to protons
How does this molecule prevent the synthesis of ATP?	Without a proton gradient, there is no energy for the F_1F_0 ATPase to utilize to combine ADP and P_i.
With the use of ETC uncouplers, energy is dissipated in what form?	Heat
What molecule prevents ATP formation by inhibiting ATP/ADP translocation in the mitochondria?	Atractyloside, thus preventing the formation of ATP

GLYCOGEN METABOLISM

What is the storage form of glucose in plants?	Starch
What breaks down starch in the body?	Starch is degraded by α-amylase in saliva and pancreatic juice to maltose, triose, and α-limit dextrans.
Where are disaccharides and monosaccharides degraded?	Surface of epithelial cells in the small intestine

By which process are monosaccharides absorbed?

Carrier-mediated transport

List the ways in which monosaccharides are utilized in the body:

Oxidized to CO_2 and H_2O for energy (via glycolysis/TCA/ETC)

Stored as glycogen

Converted to triglycerides (i.e., fat)

Released into the general circulation as glucose

In what form is glucose stored in the human body?

Glycogen

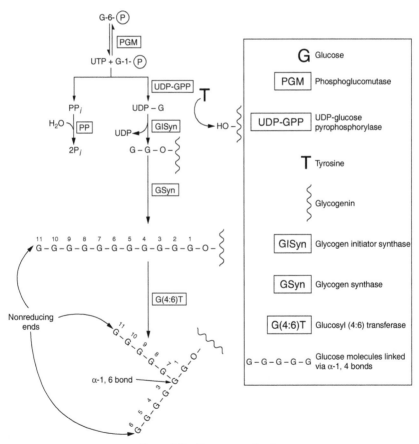

Figure 2.5   Glycogen synthesis.

Where in the body does glycogen synthesis and breakdown occur?	Liver and skeletal muscle
What two types of linkages between glucose molecules may be found in glycogen?	1. α-1,4-Glycosidic bonds (between the glucose molecules placed in a straight line) 2. α-1,6-Glycosidic bonds (placed on the branch points)
Where in the cell does glycogen synthesis take place?	Cytosol
What molecules provide the energy for the synthesis of glycogen?	ATP and uridine triphosphate (UTP)
What is the first step in the synthesis of glycogen from glucose?	Glucose-6-phosphate → glucose-1-phosphate
What enzyme catalyzes the aforementioned reaction?	Phosphoglucomutase
Uridine diphosphate (UDP)-glucose pyrophosphorylase catalyzes what reaction?	Glucose-1-phosphate + UDP → UDP-glucose
What protein serves as the primer for glycogen synthesis?	Glycogenin
What enzyme is responsible for creating the α-1,4 linkage between glucose molecules?	Glycogen synthase
Can glycogen synthase initiate glycogen synthesis de novo?	No, it can only elongate an existing glycogen chain.
Where on the newly synthesized glycogen chain are new glucose molecules added?	To the nonreducing end
What would be the shape of a glycogen molecule if glycogen synthase was the only enzyme adding glucose molecules to the chain?	Linear
What enzyme is responsible for creating the α-1,6 bonds between glucose molecules?	Glycogen-branching enzyme (GBE), also called glucosyl (α-4:6) transferase
How does this branching enzyme work?	Transfers approximately 5-8 glucosyl residues from the nonreducing end of the glycogen chain to another residue within the chain and attaches the residues via an α-1,6 linkage

What is the purpose of branching glycogen molecules?	To facilitate the breakdown of glycogen and aid in solubility
When is glycogen used as fuel?	During strenuous exercise in muscle; in the fasting state in liver
Is the process of glycogen breakdown the reverse of glycogen synthesis?	No
What enzyme is used to break down glycogen?	Glycogen phosphorylase
Where does glycogen phosphorylase cleave the glycogen molecule?	At the α-1,4 glycosidic bond
How far away from the reducing end of the glycogen molecule can glycogen phosphorylase cleave?	Discontinues cleavage when there are four glucosyl residues remaining on a chain
What structure is the product of the above cleavage reaction?	Limit dextran
Can glycogen phosphorylase further degrade this product?	No
What enzyme removes glycogen branches?	Glycogen debranching enzyme (GDE)
What are the two enzymes that constitute the debranching enzyme complex?	1. Oligo-(α-1,4 $\rightarrow$ α-1,4) glucantransferase aka glucosyl (4:4) transferase; this enzyme removes the first three glucosyl residues left on a branch of glycogen 2. Amylo-α-(1,6)-glucosidase; this enzyme removes the last glucosyl residue on a branch of glycogen
Glycogenolysis liberates what glucose-based molecule?	Glucose-1-phosphate
What enzyme converts glucose-1-phosphate to glucose-6-phosphate?	Phosphoglucomutase (reversible action)
Which enzyme found on the endoplasmic reticulum of liver cells is responsible for liberating free glucose?	Glucose-6-phosphatase
What two mechanisms regulate glycogen synthase and glycogen phosphorylase?	1. Allosteric regulation 2. Hormonal regulation
Which form of glycogen synthase, phosphorylated or nonphosphorylated, is the active form?	Nonphosphorylated

Which form of glycogen phosphorylase, phosphorylated or nonphosphorylated, is the active form?	Phosphorylated
Does glycogen synthesis occur in the fed or fasting state?	Fed state
Does glycogen breakdown occur in the fed or fasting state?	Fasting state
What hormone stimulates glycogen formation?	Insulin
What hormone stimulates glycogen breakdown?	Glucagon
In the well-fed state, what molecule allosterically activates glycogen synthase?	Glucose-6-phosphate
Name two molecules that allosterically inhibit glycogen phosphorylase:	1. Glucose-6-phosphate 2. ATP
In muscle, what effect does calcium have on glycogen phosphorylase?	Calcium binds calmodulin (a subunit of phosphorylase kinase). Phosphorylase kinase phosphorylates glycogen phosphorylase and glycogen synthase. Phosphorylated glycogen phosphorylase is activated while synthase is inactivated. Glycogen is degraded to glucose (remember that when you need a boost in the muscle or more ATP, glycogen phosphorylase is activated).
In muscle, what is the effect of cAMP on glycogen synthase and glycogen phosphorylase?	cAMP inactivates glycogen synthase; cAMP activates glycogen phosphorylase.
Describe the mechanism by which cAMP regulates glycogen synthesis:	Glucagon and epinephrine bind to their respective receptors. Adenylate cyclase is activated and produces cAMP. cAMP activates cAMP-dependent protein kinase. cAMP-dependent protein kinase phosphorylates glycogen synthase/phosphorylase kinase which then phosphorylates glycogen phosphorylase and synthase, thus activating the former and deactivating the latter enzyme. (Fig. 2.6)

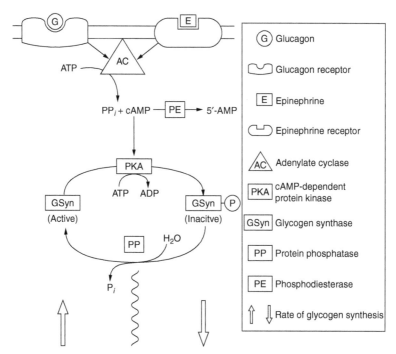

Figure 2.6 Glycogen synthesis regulation.

Name the common activators of glycogen phosphorylase:	Glucagon, epinephrine; calcium and AMP in muscle
Name the common inhibitors of glycogen phosphorylase:	Insulin, glucose, ATP, glucose-6-phosphate
Name the common activators of glycogen synthase:	Glucose-6-phosphate, insulin
Name the common inhibitors of glycogen synthase:	Glucagon, epinephrine; calcium and AMP in muscle
What is glycogen storage disease type I called?	von Gierke disease
What enzyme is deficient in von Gierke disease?	Glucose-6-phosphatase
What are the symptoms of von Gierke disease?	Hypoglycemia, hepatomegaly, renomegaly, failure to thrive, stunted growth, hyperlipidemia, hyperuricemia, 50% mortality rate

What is glycogen storage disease type II called?	Pompe disease
What enzyme is deficient in Pompe disease?	Lysosomal α-1,4 glucosidase (acid maltase)
What are the symptoms of Pompe disease?	Mild hepatomegaly, cardiomegaly, widened QRS intervals on electrocardiogram (ECG), macroglossia, muscle hypotonia, cardiorespiratory failure within 2 years of life
What is glycogen storage disease type III called?	Cori disease
What is the deficient enzyme in Cori disease?	Debranching enzyme, α-1,6-glucosidase
What are the symptoms of Cori disease?	Hepatomegaly, slow growth, low blood sugar, sometimes seizures
What is glycogen storage disease type IV called?	Andersen disease
What is the deficient enzyme in glycogen storage disease type IV?	Branching enzyme
Describe the glycogen that would be found in the cells of an individual with glycogen storage disease type IV:	Normal amount, but with very long outer branches
What are the signs and symptoms of glycogen storage disease type IV?	Liver cirrhosis with death before 2 years of age
What is glycogen storage disease type V called?	McArdle disease
What is the main organ affected in glycogen storage disease type V?	Skeletal muscle (McArdle = Muscle)
What are the symptoms of McArdle disease?	Temporary weakness and cramping of skeletal muscles after exercise, normal mental development (increased glycogen in muscle), myoglobinuria with strenuous exercise
What is the deficient enzyme in McArdle disease?	Skeletal muscle glycogen phosphorylase (remember the liver enzyme is normal)

What is glycogen storage disease type VI called?	Hers disease
What is the deficient enzyme in glycogen storage disease type VI?	Phosphorylase (liver)
What are the signs and symptoms of glycogen storage disease type VI?	Mild hepatomegaly and hyperlipidemia
What is the deficient enzyme in glycogen storage disease type VII?	PFK
Describe the symptoms of glycogen storage disease type VII:	Painful muscle cramps with exercise
What is the deficient enzyme in glycogen storage disease type VIII?	Phosphorylase kinase (liver)
What are the symptoms of glycogen storage disease type VIII?	Mild hepatomegaly and hypoglycemia
All of the glycogen storage diseases show what inheritance pattern?	Autosomal recessive

DISACCHARIDE METABOLISM

What enzymatic reaction does sucrose catalyze?	Sucrose → glucose + fructose
Where does the majority of fructose metabolism occur?	Liver
What is the first step in fructose metabolism?	Hexokinase converting fructose to fructose-6-phosphate in the muscle and kidney
Essential fructosuria results from a deficiency in what enzyme?	Fructokinase (which is liver specific)
What are the symptoms of fructosuria?	No symptoms, although fructose is seen in the blood and urine.
What reaction does aldolase b (fructose-1-phosphate aldolase) catalyze?	Fructose-1-phosphate → dihydroxyacetone phosphate (DHAP) + glyceraldehydes

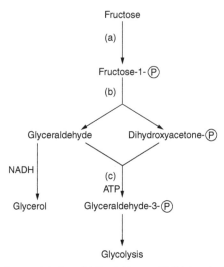

Figure 2.7   Fructose metabolism. (a) Fructokinase; (b) Aldolase b; (c) Triose kinase.

Hereditary fructose intolerance results from a deficiency in what enzyme?	Aldolase b
Describe the pathophysiology behind fructose intolerance:	The deficiency in aldolase b causes fructose-1-phosphate accumulation, thus sequestering phosphate molecules which results in decreased available phosphate. This decrease causes inhibition of glycogenolysis and gluconeogenesis.
What are the symptoms of fructose intolerance?	Jaundice, cirrhosis, hypoglycemia
How is hereditary fructose intolerance treated?	With a fructose (and sucrose)-restricted diet
What are the possible fates of glyceraldehydes?	Converted to glyceraldehyde-3 phosphate (an intermediate in glycolysis); reduced to glycerol (to be used in fatty acid synthesis and gluconeogenesis)
What reaction does lactase catalyze?	Lactose → glucose + galactose
What is the most common type of food intolerance?	Lactase deficiency

Where is the main site of lactase's action in the gastrointestinal (GI) tract?	Small intestine
What are the symptoms of congenital lactose intolerance?	Explosive, frothy stools, abdominal distention in infants exposed to milk or milk products; most experience diarrhea and malabsorption
What is the primary dietary source of galactose?	Milk
Treatment of lactase deficiency with dietary lactose avoidance necessitates the supplementation of what electrolyte into the diet?	Calcium
What is the first step in galactose metabolism?	Galactokinase converting galactose to galactose-1-phosphate
What is the main purpose of the aforementioned step?	To sequester galactose within the cell because the addition of a charged phosphate group to a molecule traps it within cells
What reaction does galactose-1-phosphate uridyltransferase catalyze?	It is involved in the conversion of galactose-1-phosphate and UDP-glucose to glucose-1-phosphate and UDP-galactose.
UDP-galactose epimerase converts UDP-galactose to what?	UDP-glucose (which cycles back for the above reaction)

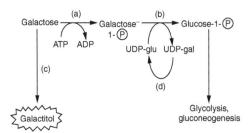

Figure 2.8 Galactose metabolism. (a) Galactokinase; (b) Uridyltransferase; (c) Aldose reductase; (d) 4-Epimerase.

Is galactokinase deficiency or galactosemia associated with a more severe clinical course?	Galactosemia

List the signs and symptoms of galactosemia:	Infants fail to thrive at birth and develop vomiting and diarrhea within a few days of milk ingestion. The most severely affected organs are the liver, brain, and eyes. Symptoms include hepatosplenomegaly, cirrhosis, cataracts, and mental retardation.
What enzyme disturbance is associated with galactosemia?	Absence of galactose-1-phosphate uridyltransferase
What inheritance pattern is seen in galactosemia?	Autosomal recessive
What are the symptoms of galactokinase deficiency?	Galactosemia in the blood and galactosuria in the urine; galactitol (a toxic metabolite of galactose metabolism) can accumulate if galactose is present in the diet; symptoms can include early congenital cataracts
What toxic metabolites are responsible for the clinical manifestations of the aforementioned disorders?	Galactitol and other galactose metabolites
Production of these toxic metabolites is the result of ectopic activity of what enzyme?	Aldose reductase
What is the definitive treatment for these disorders?	Removal of galactose (and lactose) from the diet

HEXOSE MONOPHOSPHATE SHUNT

What are the substrates for the pentose/hexose phosphate or hexose monophosphate (HMP) shunt?	Glucose-6-phosphate and nicotinamide adenine dinucleotide phosphate ($NADP^+$)
Where in the cell does the HMP shunt occur?	Cytoplasm
What organs in the body have extensive HMP shunt activity?	Lactating mammary glands, liver, adrenal cortex (erythrocytes also have HMP shunt activity)
What characteristic do these organs share?	Sites of fatty acid or steroid synthesis

What is the rate-limiting enzyme in the HMP shunt?	Glucose-6-phosphate dehydrogenase (G6PD)
Name an activator of the G6PD enzyme:	NADP$^+$
Name an inducer of the G6PD enzyme:	Insulin
Name an inhibitor of the G6PD enzyme:	NADPH
What are the major products of the HMP shunt?	Ribose-5-phosphate and NADPH
In what process is ribose-5-phosphate utilized?	In nucleotide synthesis
List three important processes which utilize NADPH:	1. Anabolic processes (as a source of reducing equivalents) 2. Respiratory/oxidative burst—rapid release of reactive oxygen species by cells (e.g., by immune cells to fight bacteria) 3. Hepatic P-450 function
How do RBCs utilize NADPH?	Via the glutathione reductase enzyme to reduce glutathione
List four important processes which utilize glutathione:	1. Reduction of protein sulfhydryl groups 2. Reduction of peroxidases 3. Maintenance of reduced hemoglobin (Hgb) 4. "Catching" amino acids in the extracellular space (via γ-glutamyltranspeptidase)

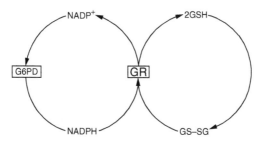

Figure 2.9 Glucose-6-phosphate dehydrogenase pathway; G6PD glucose-6-phosphate dehydrogenase, GR glutathione reductase.

What steady-state cytoplasmic NADP$^+$/NADPH ratio favors redox reactions?	1/10
Why are RBCs particularly vulnerable to oxidative damage?	RBCs have the capacity to carry large amounts of O_2 and are thus prone to oxidative damage because they have no ETC to reduce O_2.
What type of oxidative damage can occur to the hemoglobin in RBCs?	In the presence of reaction oxygen species (ROS), hemoglobin may precipitate to form Heinz bodies.
Heinz bodies are the histological hallmark of what disorder?	G6PD deficiency
What type of oxidative damage can occur to the plasma membranes of G6PD-deficient RBCs?	Peroxidation → membrane weakness → hemolytic anemia
What is the pathophysiology behind the symptoms of G6PD deficiency?	With deficient G6PD, NADPH is not regenerated, which leads to a decrease in NADPH and increase in NADP$^+$. Thus, less reduced glutathione is available to detoxify free radicals and peroxides. This results in the body's RBCs having a poorer defense against oxidizing agents, which leads to hemolytic anemia.
In what situations would oxidative damage in the presence of G6PD deficiency be accelerated?	Ingestion of foods containing oxidants (i.e., fava beans) Treatment with certain drugs (i.e., sulfonamides, antituberculosis drugs) Infections (i.e., pneumonia, infectious hepatitis)
In what ethnic group is G6PD deficiency most prevalent?	African Americans
Are males or females more likely to present with G6PD deficiency?	Males (X-linked recessive disorder)
How do polymorphonuclear leukocytes (PMNs) utilize NADPH?	NADPH is used by NADPH oxidase to produce ROS that destroy bacteria.
What is the most common cause of chronic granulomatous disease (CGD)?	NADPH oxidase deficiency in PMNs

What are the clinical manifestations of CGD?	Increased susceptibility to infection by catalase-positive organisms such as *Escherichia coli, Staphylococcus aureus,* and *Klebsiella*; increased risk of lymphoma
What test is used to confirm the diagnosis of CGD?	A negative nitroblue tetrazolium test
How do hepatocytes utilize NADPH?	In the biosynthesis of fatty acids, cholesterol, and nucleotides
Describe the pathogenesis of alcohol-induced hepatic steatosis:	Alcohol metabolism disrupts the NADH/NAD$^+$ ratio which prevents the utilization of glycerol-3-phosphate as a gluconeogenic substrate while increasing its diversion into triglycerides, resulting in hepatic steatosis.

HEME METABOLISM

Name the two major sites of heme synthesis:	1. Bone marrow 2. Liver
What enzyme catalyzes the rate-limiting step of heme synthesis?	Aminolevulinate (ALA) synthase
What reaction does this enzyme catalyze?	Glycine + succinyl CoA → δ-aminolevulinate

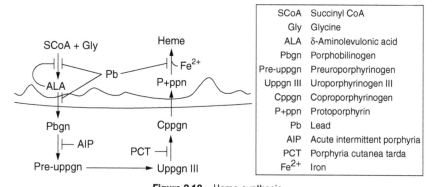

SCoA	Succinyl CoA
Gly	Glycine
ALA	δ-Aminolevulonic acid
Pbgn	Porphobilinogen
Pre-uppgn	Preuroporphyrinogen
Uppgn III	Uroporphyrinogen III
Cppgn	Coproporphyrinogen
P+ppn	Protoporphyrin
Pb	Lead
AIP	Acute intermittent porphyria
PCT	Porphyria cutanea tarda
Fe^{2+}	Iron

Figure 2.10 Heme synthesis.

In what cellular compartment is this enzyme found?	Mitochondria
What molecule is the major inhibitor of this enzyme?	Heme (feedback inhibition)
Underproduction of heme results in what type of anemia?	Microcytic, hypochromic anemia
Define porphyria:	A group of disorders involving heme biosynthesis, characterized by the excessive excretion of porphyrins or their precursors
Describe the mechanism behind lead poisoning:	Lead inhibits ferrochelatase and ALA synthase
List six clinical features of lead poisoning:	1. Headache 2. Nausea 3. Abdominal pain 4. Memory loss 5. Neuropathy 6. Lead lines in gums
What characteristic histological feature on a peripheral blood smear is found with lead poisoning?	Coarse basophilic stippling of erythrocytes
What two molecules accumulate in the urine in the setting of lead poisoning?	1. Coproporphyrin 2. ALA (urine may appear pink in color)
What is the enzymatic defect in acute intermittent porphyria (AIP)?	Deficiency in uroporphyrinogen I synthetase
What two molecules accumulate in the urine in AIP?	1. Porphobilinogen 2. δ-ALA
Name three features of AIP:	1. Anxiety 2. Abdominal pain 3. Autosomal dominant inheritance
What is the enzymatic defect in porphyria cutanea tarda (PCT)?	Deficiency of uroporphyrinogen decarboxylase
What molecules accumulate in the urine in PCT?	Uroporphyrin accumulates in urine, resulting in a tea color.
What is the most common clinical feature of PCT?	Photosensitivity, resulting in inflammation and blistering of skin.

Describe the production of bilirubin: Heme is recovered from hemoglobin after hemolysis of RBCs in the spleen. This is subsequently converted to biliverdin and then bilirubin. (Fig. 2.11)

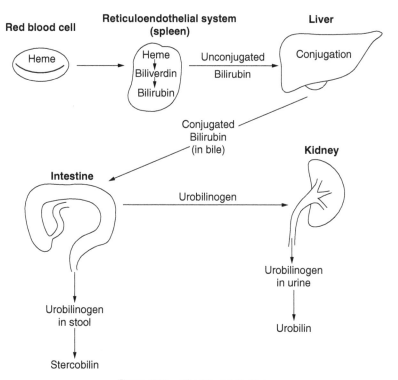

Figure 2.11 Bilirubin metabolism.

How is unconjugated bilirubin transported in the blood? Via albumin because bilirubin is not water soluble

What role do hepatocytes play in the metabolism of bilirubin? Hepatocytes conjugate bilirubin with glucuronate via UDP-glucuronyl transferase, thus increasing its water solubility.

How does bilirubin enter the intestine? Following conjugation, bilirubin is secreted into bile.

How is bilirubin processed in the intestine? Intestinal bacteria convert conjugated bilirubin into urobilinogen.

How is bilirubin excreted in feces?

A portion of urobilinogen is converted into bile pigments (stercobilin) and excreted in feces.

How is bilirubin excreted in urine?

A portion of urobilinogen is converted into urobilin (yellow) and is excreted in urine.

What is jaundice?

A yellowish staining of the integument, sclerae, and deeper tissues and the excretions with bile pigments

Describe the types of hyperbilirubinemia, urine bilirubin, and urine urobilinogen levels in the following types of jaundice:

Hepatocellular

Conjugated (direct) or unconjugated (indirect) hyperbilirubinemia, increased urine bilirubin, normal or decreased urine urobilinogen

Obstructive

Conjugated (direct) hyperbilirubinemia, increased urine bilirubin, decreased urine urobilinogen

Hemolytic

Unconjugated (indirect) hyperbilirubinemia, decreased/absent urine bilirubin, increased urine urobilinogen

Name some hereditary hyperbilirubinemias:

Gilbert syndrome, Crigler-Najjar syndrome, Dubin-Johnson syndrome, Rotor syndrome, biliary atresia, and physiologic jaundice in the newborn

Which hereditary hyperbilirubinemias result from dysfunctional UDP-glucuronyl transferase?

Gilbert syndrome, Crigler-Najjar syndrome

What type of hyperbilirubinemia is associated with these disorders?

Unconjugated hyperbilirubinemia

Which of the aforementioned disorders is typically asymptomatic?

Gilbert syndrome (decreased UDP-glucuronyl transferase activity)

Which form of Crigler-Najjar syndrome is typically lethal early in life?

Type I

What are the signs and symptoms of this form of Crigler-Najjar syndrome?

Jaundice, kernicterus, increased unconjugated bilirubin

What is the treatment of this form of Crigler-Najjar syndrome?	Plasmapheresis, phototherapy
Which form of Crigler-Najjar syndrome responds to phenobarbital treatment?	Type II
Describe the pathophysiology behind Dubin-Johnson syndrome:	Defective liver excretion of conjugated bilirubin
On autopsy, will an individual who suffered from Dubin-Johnson or Rotor syndrome have a blackened liver?	Dubin-Johnson syndrome results in blackened liver
What type of hyperbilirubinemia occurs with viral hepatitis or cirrhosis?	Conjugated and unconjugated hyperbilirubinemia, as well as increased aminotransferase (alanine aminotransferase/aspartate aminotransferase [AST/ALT]) levels

PURINE AND PYRIMIDINE METABOLISM

Describe the structure of a nucleoside:	A nitrogenous base linked to a pentose monosaccharide
What differentiates a nucleoside from a nucleotide?	Addition of a phosphate group(s)
What bases make up the purine nucleotides?	Adenosine (A) and guanine (G)
What bases make up the pyrimidine nucleotides?	Cytosine (C), uracil (U), and thymine (T)
What compounds constitute the purine ring?	Amino acids (i.e., aspartic acid, glycine, glutamine), CO_2, and N^{10}-formyl-tetrahydrofolate
What enzyme catalyzes the rate-limiting step of the purine synthesis pathway?	Phosphoribosylpyrophosphate (PRPP) synthetase
What are the inhibitors of PRPP synthetase?	Inosine monophosphate (IMP), AMP, and guanosine monophosphate (GMP)
What is the rate-limiting step of the purine pathway?	Synthesis of 5-phosphoribosyl-1-pyrophosphate from ribose-5-phosphate

What is the committed step of the purine pathway?	Formation of 5'-phosphoribosylamine from 5-phosphoribosyl-1-pyrophosphate
The next several steps in the de novo purine synthesis pathway ultimately form what compound?	IMP
How many ATP molecules does this require?	Four
Upon synthesis of IMP, there is a fork in the pathway that leads to the formation of what two molecules?	1. AMP 2. GMP
What molecule is the energy source for AMP formation?	GTP
What molecule is the energy source for GMP formation?	ATP
Are AMP and GMP positive or negative regulators of their own synthesis?	Negative regulators
What enzymes are involved in the conversion of nucleoside monophosphates into nucleoside di- or triphosphates?	Nucleoside monophosphate kinases such as adenylate kinase and guanylate kinase, and nucleoside diphosphate kinases
What enzyme catalyzes the following reaction: AMP + ATP → ADP?	Adenylate kinase
What enzyme catalyzes the following reaction: GMP + ATP → GDP + ADP?	Guanylate kinase
Describe the reactions that nucleoside diphosphate kinase catalyze:	Interconversion of nucleoside diphosphates and triphosphates (i.e., GDP + ATP → GTP + ADP or CDP + ATP → CTP + ADP)
What drugs act to inhibit the conversion of N^{10}-formyl-tetrahydrofolate to tetrahydrofolate?	p-Aminobenzoic acid (PABA) antimetabolites (i.e., sulfonamides) and folic acid analogs (i.e., methotrexate)
What is the mechanism of action of PABA analogs?	Competitively inhibit dihydropteroate synthase, which leads to decreased bacterial synthesis of folic acid
Do these drugs affect human purine synthesis?	No, because humans cannot make folic acid.

PABA analogs can be used to treat what bacterial infections?

Gram-positive, gram-negative, *Nocardia, Chlamydia*

What is the mechanism of action of folic acid analogs?

Competitively inhibit dihydrofolate reductase, resulting in decreased deoxythymidine monophosphate (dTMP), and thus decreased DNA and protein synthesis

What drug may be given to prevent toxicity from folic acid analog use?

Leucovorin (folinic acid)

How does this drug work?

Leucovorin (folinic acid) can be readily converted to tetrahydrofolate but does not require dihydrofolate reductase for its conversion.

What is the function of the purine salvage pathway?

To convert purines that result from cell turnover or from the diet; those that cannot be degraded are turned into nucleoside triphosphates

What are the three main enzymes involved in the purine salvage pathway?

Xanthine oxidase, HGPRT (hypoxanthine phosphoribosyl-transferase), and adenosine deaminase (ADA)

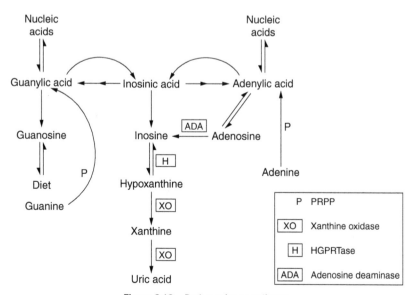

Figure 2.12 Purine salvage pathway.

What disease is the result of ADA deficiency?	Severe combined immunodeficiency (i.e., SCID, an immunodeficiency in both B and T cells)
How does ADA deficiency cause this disease?	ADA deficiency causes an excess of ATP and dATP; this impacts negatively on ribonucleotide reductase and thus prevents DNA synthesis, which decreases B- and T-lymphocyte count.
Does this disease most commonly present in childhood or adulthood?	Childhood
What types of infections are individuals with this disease likely to suffer from?	Recurrent viral, bacterial, fungal, and protozoal infections
What disease is the result of HGPRT deficiency?	Lesch-Nyhan syndrome
What are some findings associated with this disorder?	Mental retardation, aggression, self-mutilation, hyperuricemia, gout, and choreoathetosis
Serum levels of what substance are increased in Lesch-Nyhan syndrome?	Uric acid
What is the inheritance pattern of Lesch-Nyhan syndrome?	X-linked recessive
What is gout?	A disorder of purine metabolism characterized by a raised but variable blood uric acid level and severe recurrent acute arthritis of sudden onset resulting from deposition of sodium urate crystals in connective tissue and articular cartilage
Where is the characteristic location of an acute attack of gout?	Metatarsophalangeal joint of the great toe
Describe the sodium urate crystals:	Needle-shaped and negatively birefringent
List the possible causes of gout:	Lesch-Nyhan syndrome PRPP excess Decreased excretion of uric acid G6PD deficiency Administration of thiazide diuretics

What is allopurinol's mechanism of action?	Xanthine oxidase inhibitor
What substances have increased levels as a result of allopurinol's mechanism of action?	Hypoxanthine and xanthine
Can these substances crystallize?	No
What is the mechanism by which colchicine treats gout?	Depolymerizes microtubules and thus reduces inflammation by impairing leukocyte chemotaxis and degranulation
What is the mechanism by which probenecid treats gout?	Probenecid inhibits the reabsorption of uric acid in the proximal tubule
Which antigout drug delivers the most immediate relief following administration?	Colchicine, since gout pain is primarily due to inflammation.
What molecules provide the source of carbon and nitrogen atoms in the pyrimidine ring?	Glutamine, aspartic acid, and CO_2
What is the committed step in pyrimidine synthesis?	Carbamoyl phosphate formation
What catalyzes this reaction?	Carbamoyl phosphate synthetase II (CPS II)
What cofactor is normally used in carboxylating reactions?	Biotin
Does the aforementioned enzyme utilize this cofactor?	No
In what other cycle is carbamoyl phosphate formed?	Urea cycle
Name the key differences between CPS I and CPS II:	CPS I is absolutely dependent upon the presence of the allosteric activator *N*-acetylglutamate.
	CPS I is involved in the urea cycle; CPS II is involved in pyrimidine synthesis.
	CPS I is in the mitochondria; CPS II is in the cytosol.
	CPS I utilizes ammonia as its source of nitrogen; CPS II utilizes the γ-amide group of glutamate as the source of nitrogen.

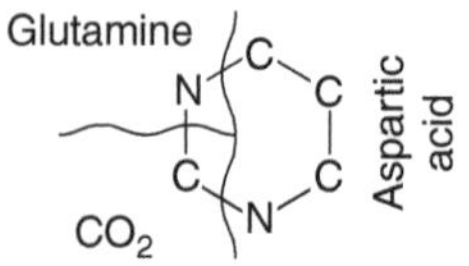

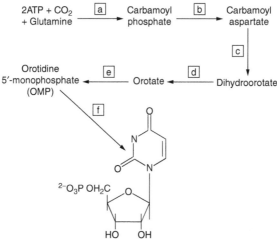

Figure 2.13   Pyrimidine constituents.

Figure 2.14   Pyrimidine synthesis. (a) Carbamoyl phosphate synthetase II; (b) Aspartate transcarbamoylase; (c) Dihydroorotase; (d) Dihydroorotate dehydrogenase; (e) Orotate phosphoribosyl-transferase; (f) OMP decarboxylase.

Uridine 5′-monophosphate (UMP) can be converted into which compound?	Cytidine monophosphate (CMP)
What is the difference between UMP and TMP?	TMP is the methylated version of UMP
Can the unmodified products of purine and pyrimidine synthesis be used for RNA synthesis?	Yes
Can the unmodified products of purine and pyrimidine synthesis be used for DNA synthesis?	No, they must be converted from ribonucleotides to deoxyribonucleotides.
What enzyme catalyzes the conversion of ribonucleotides to deoxyribonucleotides?	Ribonucleotide reductase
What molecule inhibits this enzyme?	dATP

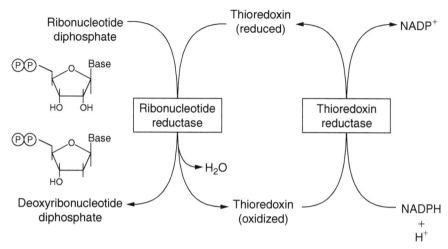

Figure 2.15 Conversion of ribonucleotides to deoxyribonucleotides.

AMINO ACID TRANSPORT

What is the first step in the breakdown of amino acids?	Removal of the α-amino group, usually transferring this group to α-ketogluatarate
What enzymes participate in the transfer of α-amino groups?	Aminotransferases such as ALT and AST
Elevations in the plasma levels of the aforementioned enzymes are correlated with disease in what organ?	Liver
What step follows glutamate formation?	Oxidative deamination of glutamate
What enzyme catalyzes this reaction?	Glutamate dehydrogenase

| What coenzymes does this enzyme utilize? | NAD^+ and $NADP^+$ |

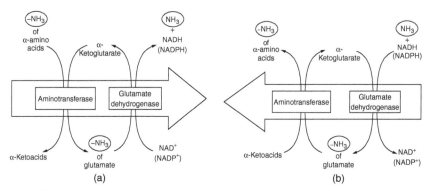

Figure 2.16   Disposal and synthesis of amino acids. (a) Disposal; (b) Synthesis.

UREA CYCLE

What is the function of urea?	To dispose of amino groups from amino acids, because ammonia is toxic to the body (particularly the central nervous system [CNS])
What molecules provide the two nitrogen atoms of urea?	One nitrogen comes from free NH_3 and the other nitrogen comes from aspartate (the carbon and oxygen of urea comes from CO_2).
Where in the body does the urea cycle take place?	Liver
Where in the cell does the urea cycle take place?	The first two reactions take place in the mitochondria and the rest of the cycle takes place in the cytosol.
What two molecules are able to cross the mitochondrial membrane?	1. Ornithine 2. Citrulline
What is the rate-limiting step of the urea cycle?	Formation of carbamoyl phosphate

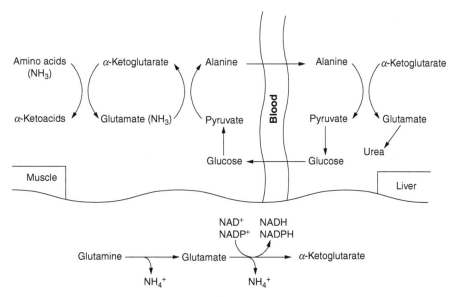

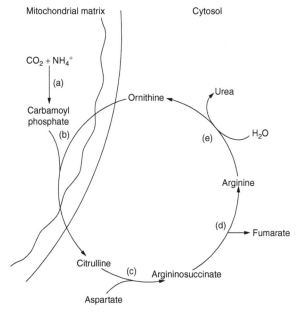

Figure 2.17 Ammonium transport by alanine and glutamine.

Figure 2.18 Urea cycle. (a) Carbamoyl phosphate synthase; (b) Ornithine transcarbamoylase; (c) Argininosuccinate synthase; (d) Aminotransferase; (e) Arginase.

What enzyme catalyzes the rate-limiting step of the urea cycle?	CPS I
How many phosphate groups are used in the formation of urea?	Four
Where in the body is urea transported for excretion?	Kidneys

CHOLESTEROL, LIPOPROTEINS, AND STEROID BIOSYNTHESIS

What molecule provides the carbon pool for steroid synthesis?	Acetate
What organ plays the most significant role in cholesterol balance?	Liver
Which molecule provides the reducing equivalents for cholesterol synthesis?	NADPH
What enzyme catalyzes the rate-limiting step in the synthesis of cholesterol?	Hydroxymethylglutaryl (HMG)-CoA reductase
What series of reactions make the synthesis of cholesterol essentially irreversible?	The four condensation reactions which release pyrophosphate
Which molecules provide feedback inhibition to HMG-CoA reductase?	Cholesterol, cAMP
Name a hormone which decreases the rate of cholesterol synthesis:	Glucagon
How does this type of negative regulation occur?	Glucagon favors formation of the phosphorylated (inactive) form of HMG-CoA reductase.
Name a hormone which increases the rate of cholesterol synthesis:	Insulin
How does this type of negative regulation occur?	Insulin favors the formation of the unphosphorylated (active) form of HMG-CoA reductase.

How do sterols inhibit de novo cholesterol synthesis?

Sterols interact with SCAP, which in turn regulates the activity of the transcription factor SREBP and thus the transcription of HMG-CoA reductase (HMGR). HMGR also contains a sterol-sensing domain and when sterols are high the protein is targeted for ubiquitination and degradation in the proteosome.

Name three organs that convert cholesterol into steroid hormones:

1. Adrenal cortex
2. Gonads (testes and ovaries)
3. Placenta

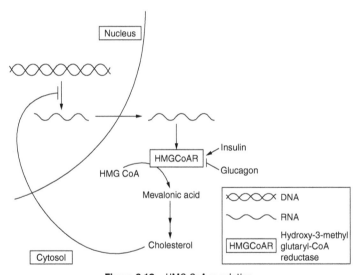

Figure 2.19   HMG CoA regulation.

What is the mechanism of action of statins?

Inhibit HMG-CoA reductase and decrease the rate of de novo cholesterol synthesis

What is the composition of a lipoprotein?

Varying proportions of cholesterol, triglycerides, and phospholipids in addition to associated apolipoproteins

Name the four major apolipoproteins and list their functions:

1. A-I activates lechitin-cholesterol acyltransferase.
2. B-100 binds the LDL receptor.
3. C-II cofactor for lipoprotein lipase.
4. E functions with B-100 in LDL receptor interaction, binds the apo E receptor.

What is the function of a chylomicron?

To deliver dietary triglycerides to peripheral tissues and dietary cholesterol to the liver

What cells make chylomicrons?

Enterocytes

What is the function of very low-density lipoproteins (VLDLs)?

To deliver hepatic triglycerides to peripheral tissues

From where are VLDLs secreted?

Liver

What is the function of LDLs?

To deliver hepatic cholesterol to peripheral tissues

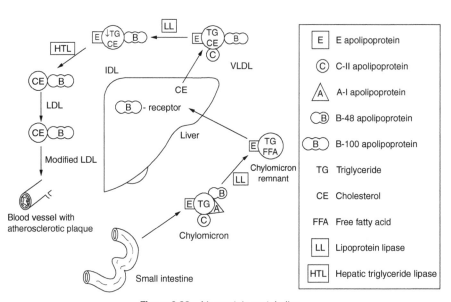

Figure 2.20   Lipoprotein metabolism.

How are LDLs formed?	Via the modification of VLDLs by lipoprotein lipase in peripheral tissues
How are LDLs taken up into target cells?	Receptor-mediated endocytosis
What is the pathophysiology behind familial hypercholesterolemia?	Increased LDL (bad cholesterol) due to a defective LDL receptor
What total cholesterol level can be found in individuals heterozygous for the aforementioned mutation?	About 300 mg/dL
What total cholesterol level can be found in individuals homozygous for the aforementioned mutation?	About 700 mg/dL
What are the symptoms of familial hypercholesterolemia type IIa?	Severe atherosclerosis early in life, possible myocardial infarction before 20 years of age, xanthomas
Familial hypercholesterolemia type IIa is inherited in what pattern?	Autosomal dominant
From where are high-density lipoproteins (HDLs) secreted?	Liver and intestine
What is the function of an HDL?	To mediate transport of cholesterol from peripheral tissues to the liver
Intermediate density lipoproteins (IDLs) are formed from the degradation of what lipoproteins?	VLDLs
What is the function of an IDL?	To deliver triglycerides and cholesterol to the liver
What is the structure of triglycerides?	Three fatty acids esterified to a glycerol backbone
What enzymes hydrolyze triglycerides?	Pancreatic lipase, lipoprotein lipase, hormone-sensitive lipase
What substance secreted by the liver aids in the digestion of fatty acids?	Bile
Where is bile stored?	Gall bladder

Bile duct obstruction can lead to a deficiency in what vitamins?	Vitamins A, D, E, and K (fat soluble)
Why does bile duct obstruction lead to this deficiency?	Absorption of these vitamins is dependent on the presence of bile
Following the absorption of triglycerides by the epithelial cells of the small intestine, what are triglycerides combined to form?	Chylomicrons
What must happen to a chylomicron in order for it to be taken up by tissue?	Hydrolysis by clearing factor lipase (i.e., lipoprotein lipase)
Lipoprotein lipase hydrolyzes triacylglycerides into what molecules?	2-Monoacylglycerol and fatty acids
Triglyceride is transported from the liver to adipose tissue in what form?	VLDL
What enzyme is activated in the fasting state to mobilize stored triglycerides?	Hormone-sensitive lipase
Name the four principal functions of fatty acids:	1. Components of phospholipids 2. Lipophilic modifiers of proteins 3. Fuel molecules 4. Hormones and intracellular messengers
Name two types of cells which cannot utilize fatty acids as a form of energy:	1. Erythrocytes 2. Brain cells
What molecule, which can replace glucose as a fuel, can fatty acids be converted into?	Ketone bodies (excess ketones produced in the body can sometimes result in ketonuria, ketones in the urine)
In what states are high levels of ketone bodies observed?	Fasting state (insufficient glucose) and diabetes (dysfunctional glucose uptake secondary to insufficient insulin or insulin insensitivity)
Glycerol is utilized by what tissue?	Liver
What is glycerol converted to in this tissue?	Dihydroxyacetone phosphate (i.e., DHAP, which can be converted to glucose)

What two enzymes transport fatty acids across the mitochondrial membrane?	1. Carnitine palmitoyltransferase I 2. Carnitine palmitoyltransferase II (see Fig. 2.21)

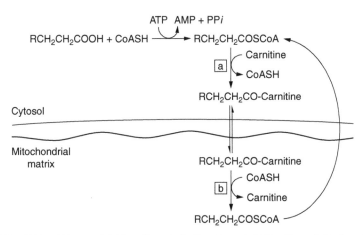

Figure 2.21 Fatty acid transport. (a) Carnitine palmitoyltransferase I; (b) Carnitine palmitoyltransferase II.

Defects in the carnitine transport system can lead to what clinical manifestations?	Hypoglycemia, muscle wasting
What is the name of the process that oxidizes fatty acids to acetyl CoA?	β-Oxidation
Where in the cell does this process occur?	Mitochondrial matrix
How does malonyl CoA inhibit the oxidation of fatty acids?	Inhibits carnitine palmitoyltransferase I
What are the final products in the oxidation of fatty acids with odd numbers of carbon atoms?	Propionyl CoA and acetyl-CoA
This product must be converted into what molecule in order to enter the TCA cycle?	Succinyl CoA
What reaction does propionyl-CoA carboxylase catalyze?	Propionyl CoA to methylmalonyl CoA
What is the pathophysiology behind propionyl-CoA carboxylase deficiency?	Deficiency of propionyl-CoA carboxylase causes odd-numbered fatty acid chains to build up and accumulate in the liver.

What are the symptoms of propionyl-CoA carboxylase deficiency?   Episodic lethargy, anorexia, vomiting, acidosis, CNS depression, and developmental problems

What inheritance pattern is seen in propionyl-CoA carboxylase deficiency?   Autosomal recessive

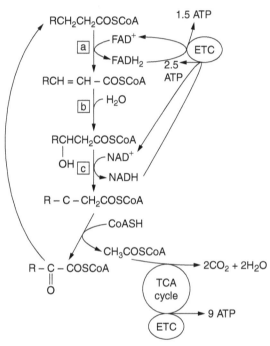

Figure 2.22   Pathway for β-oxidation. (a) Acyl-CoA dehydrogenase; (b) Enoyl hydratase; (c) β-hydroxyacyl-CoA dehydrogenase.

Name the ketone bodies: Acetoacetate, β-hydroxybutyrate, acetone

What diseases predispose to ketoacidosis?   Ketoacidosis is most common in untreated type 1 diabetes mellitus; prolonged alcoholism may lead to alcoholic ketoacidosis.

What two characteristics of ketone bodies make them a good fuel for the glucose-starved brain?
1. Free solubility in blood
2. Easy crossing of the blood-brain barrier

What two tissues synthesize fatty acids?
1. Adipose tissue
2. Liver

Where in the cell does fatty acid synthesis occur?	Cytosol
What is the principal regulated reaction in fatty acid synthesis?	Acetyl CoA + CO_2 + ATP → malonyl CoA + ADP + P_i
What enzyme catalyzes this reaction?	Acetyl-CoA carboxylase
What are the main hormone regulators of this enzyme?	Stimulated by insulin; inhibited by glucagon and epinephrine
Name two allosteric regulators of this enzyme:	1. Citrate (+) 2. Palmitoyl CoA (−)

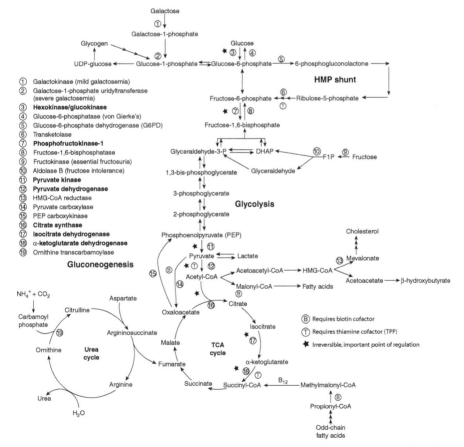

Figure 2.23 Summary of the metabolic pathways. (Reproduced with permission from Le, T. et al. *First aid for the USMLE Step 1,* New York: McGraw Hill, 2009, p100.)

CLINICAL VIGNETTES—MAKE THE DIAGNOSIS

A 32-year-old white male is recovering from what was properly diagnosed as an acute myocardial infarction the previous day. Physical examination reveals xanthelasmas, arcus senilis, and painful xanthomas of the Achilles tendon and patellae. The man's father died of a myocardial infarction, before he was 40 years of age. Serum levels of low-density lipoprotein (LDL) are extremely high.

This patient likely suffers from what disorder?	Type II hyperlipoproteinemia; familial hypercholesterolemia
What is the inheritance pattern of this disorder?	Autosomal dominant
What are the potential treatments of this disorder?	Meticulous dieting, cholesterol-lowering drugs

A 17-year-old white female complains of midepigastric pain and nausea after eating fried foods. She also has an older sibling who suffers from the same symptoms. Her face, scalp, elbows, and knees have nonpainful, yellowish papules and she has marked hepatosplenomegaly. Laboratory tests reveal very high triglyceride levels and moderate elevation of serum cholesterol and phospholipids.

This girl's complaints and family history are suggestive of what diagnosis?	Familial hypertriglyceridemia (autosomal dominant)
What additional markers may be elevated in this girl?	Serum amylase and lipase (recurrent acute pancreatitis)
What are potential treatments of this disorder?	Low-fat diet, exercise, avoidance of alcohol

A 21-year-old gentleman is being evaluated for progressive muscle weakness. He is unable to raise his arms above his shoulders and anytime he stands for longer than 2 hours, he experiences severe pain in his legs. Electromyogram (EMG) studies were unrevealing and so a muscle biopsy was done, which reveals extensive accumulation of membrane-bound glycogen along with absence of myofilaments and sarcoplasmic organelles.

This patient is likely deficient in what enzyme?	α-1,4 glucosidase (Pompe disease)
Which organs are most significantly involved?	Heart and skeletal muscle
Diagnosis can be confirmed by what method?	Enzyme assay on leukocytes or fibroblasts

A 3-year-old boy is being evaluated for recurrent infections. He is found to be leukopenic and has megaloblastic hypochromic anemia. He is also noted for developmental retardation. Over the next couple of months, his anemia is found to be unresponsive to iron, folic acid, or vitamin B_{12}. High levels of orotic acid are found in his urine.

What metabolic defect could cause this?	A defect in pyrimidine metabolism
What is the inheritance pattern of this disease?	Autosomal recessive
What is a possible treatment for this disease?	High-dose oral uridine

A 2-month-old girl is brought to the emergency room because her mother had difficulty awakening her from an overnight sleep. The baby is found to be extremely lethargic. Laboratory workup reveals a very low blood glucose level and elevated lactate with a large anion gap. Glucagon administration produced a marked increase in lactate without hyperglycemia. Physical examination revealed a markedly enlarged liver and nonpalpable spleen.

This patient is deficient in what enzyme?	Glucose-6-phosphatase (von Gierke disease)
Besides hypoglycemia, what is another metabolic consequence of this disease?	Hyperlipidemia
How can this disease be treated?	Nasogastric feeding to maintain glucose levels for infants, raw cornstarch can be used for older children

A 6-year-old Caucasian male is being evaluated for anemia. Past medical history was significant for severe jaundice and anemia at birth which was treated with an exchange transfusion. Physical examination revealed mild splenomegaly. Hemoglobin electrophoresis revealed normal hemoglobin. Erythrocyte osmotic fragility was normal and the Coombs test was negative.

What enzyme deficiency could cause this?	Pyruvate kinase deficiency
Patients are usually asymptomatic in nonsevere forms. What could cause symptoms in nonsevere forms?	Aplastic crisis
What viral infection is associated with the above?	Parvovirus B19

A 55-year-old man is seen in the emergency room complaining of severe pain in his right first metatarsophalangeal joint. He reports binge drinking after a fight with his wife earlier that day. His great toe is very painful to active and passive motion. In addition, physical examination reveals a lump under the skin on his left ear and olecranon bursitis.

This patient is likely suffering from what disorder?	Gout
What would be expected to be seen on aspiration of the synovial fluid in the involved joint?	Negatively birefringent, needle-shaped crystals of uric acid salts
What are potential acute and long-term treatments of this disorder?	Colchicine and nonsteroidal anti-inflammatory drugs (NSAIDs) acutely; allopurinol and probenecid long term

A 20-year-old college student presents in the emergency room seemingly confused with complaints of abdominal pain, diarrhea, and vomiting. While she is talking, a peculiar fruity breath smell is noted. She reports tailgating all day and being intoxicated following a party earlier that night. She cannot remember whether she had taken her diabetes medication.

This patient is most likely what type of diabetic?	Insulin-dependent diabetes mellitus, type I, juvenile onset
How would her glucose, bicarbonate, and anion gap levels be changed?	Increased glucose, decreased bicarbonate, increased anion gap
What is the proper order of treatment?	Correction of fluid deficit if dehydrated, potassium administration, gradual lowering of glucose with insulin

An 8-week-old white male is brought into the pediatrician's office because of lethargy, difficulty feeding, occasional vomiting, and yellowing of the skin. The patient is growing in the fifth percentile. On physical examination, the patient shows irritability, jaundice, hepatomegaly, and bilateral cataracts. Urine analysis shows galactosuria.

This patient is likely suffering from what disorder?	Galactosemia
What enzyme is deficient in this patient?	Galactose-1-phosphate uridyl transferase
Which organs are most severely affected by this disorder?	Liver, eyes, and brain (deposition of galactose-1-phosphate and galactitol)

A 6-month-old baby is brought to her pediatrician because the parents have noticed frequent nausea, vomiting, and lethargy following the addition of fruit juices to the baby's previous diet of only breast milk. Laboratory tests reveal hypoglycemia with fructosemia and the baby's urine is positive for reducing sugars.

What enzyme is deficient in this baby?	Aldolase b (hereditary fructose intolerance)
If untreated, this disease may lead to what complication?	Liver cirrhosis
What is the treatment for this disease?	Avoiding fructose in the diet (i.e., fruit juices, fruits, sweets)

An 18-year-old black male enlisted in the military service is vaccinated before being shipped off to central Africa. Several days later, he develops a fever and complains of weakness and fatigue. Physical examination reveals mild jaundice and slight nail bed pallor. On direct questioning, the man says that he was vaccinated against malaria amongst other illnesses.

This patient is likely deficient in what enzyme?	Glucose-6-phosphate dehydrogenase
CBC and liver function tests (LFTs) would likely show what results?	Low hemoglobin, low hematocrit, reticulocytosis, elevated direct bilirubin
What is the inheritance pattern of this disease?	X-linked recessive

CHAPTER 3

Nutrition

ENERGY NEEDS

What is the basal energy expenditure?	Energy used for metabolic processes while at rest
The basal energy expenditure represents what percentage of the total energy expenditure?	60%
What is the thermic effect of food?	Energy required for digesting and absorbing food
The thermic effect of food represents what percentage of the total energy expenditure?	10%
What is the activity-related expenditure?	Energy that varies with the level of physical activity
The activity-related expenditure represents what percentage of the total energy expenditure?	20%–30%
What is the estimated daily energy need for an infant?	120 kcal/kg dry body weight
What is the estimated daily energy need for an adult?	30 kcal/kg dry body weight
What is the caloric yield from 1 g of carbohydrate?	4 kcal
What is the caloric yield from 1 g of protein?	4 kcal
What is the caloric yield from 1 g of fat?	9 kcal
What is the caloric yield from 1 g of alcohol?	7 kcal

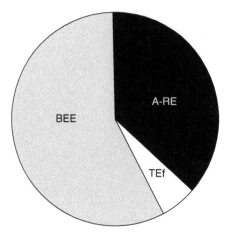

Figure 3.1   Total energy expenditure. BEE—Basal energy expenditure; TE$_f$—Thermic effect of food, A–RE—Activity-related expenditure.

MACRONUTRIENTS

Carbohydrates should typically comprise what percentage of the total caloric intake?	50%–60%
What is the difference between "available" and "unavailable" carbohydrates?	"Available" carbohydrates can be used by tissues for fuel; "unavailable" carbohydrates are not digested or absorbed but do provide bulk to the diet and assist in elimination.
List the "available" carbohydrates:	Glucose, fructose, sucrose, lactose, maltose, starches, dextrins, and glycogen
List the "unavailable" carbohydrates:	Cellulose, hemicellulose, lignin, pectins, and gums
Inadequate carbohydrate intake may lead to what catastrophic processes?	Ketosis, wasting, cationic loss, dehydration
In what forms are excess carbohydrates stored in the body?	Glycogen, triacylglycerol
Fats should typically comprise what percentage of the total caloric intake?	30% (saturated fats should make up less than 10%)
List the essential fatty acids:	Linoleic acid, linolenic acid

What functions do fats have in the body?	Precursors for synthesis of prostaglandins, prostacyclins, leukotrienes, and thromboxanes; carriers of fat-soluble vitamins; slow gastric emptying; give foods a desirable texture and taste
What characteristic symptom is associated with inadequate fat intake?	Scaly dermatitis
In what form is excess fat stored in the body?	Triacylglycerol
Proteins should typically comprise what percentage of the total caloric intake?	10%–20%
What is the recommended adult protein intake?	0.8 g/kg body weight per day
List the nine essential amino acids that cannot be synthesized in the body from nonprotein precursors:	**MILK WiTH V**eggies and **F**ruits is essential: **M**—methionine, **I**—isoleucine, **L**—leucine, **K**—lysine, **W**—tryptophan, **T**—threonine, **H**—histidine, **V**—valine, **F**—phenylalanine
What function do proteins have in the body?	Provide amino acids for synthesizing proteins and nonprotein nitrogenous bases
What is nitrogen balance?	The difference between nitrogen intake (protein) and nitrogen excretion (undigested protein in the feces + urea in the urine + ammonia in the urine)
In what instances would an individual be in positive nitrogen balance?	Pregnancy/lactation, growth, recovery from trauma/infection/surgery
In what instances would an individual be in negative nitrogen balance?	Metabolic stress, insufficient dietary protein, insufficient intake of an essential amino acid
What is kwashiorkor?	A form of malnutrition caused by inadequate protein intake in the presence of adequate total caloric intake (often occurs in areas of limited food/protein supply and/or inadequate knowledge of proper diet)

What are the clinical features of kwashiorkor?	Protuberant belly, fatigue, irritability, growth failure, loss of muscle mass, edema; skin disorders such as vitiligo, alopecia, and dermatitis are also common; severe, late-stage protein deficiency can lead to shock, coma, or death
What is marasmus?	Protein-energy malnutrition that results from a negative energy balance (decreased total caloric intake, increased energy expenditure, or a combination of both)
What are the clinical features of marasmus?	Children adapt to the energy deficit with a decrease in physical activity, lethargy, slowing of growth, muscle wasting, and loss of subcutaneous fat; other clinical features include anemia, edema, tachypnea, and abdominal distention.
What equation calculates BMI (body mass index)?	Weight (kg)/Height (m^2) = BMI (kg/m^2)
What is an ideal BMI?	19–25
What BMI is correlated with obesity for women? For men?	Over 32; over 31
What is the correlation between BMI and poor health?	The risk of poor health increases with increasing BMI.
List the diseases associated with obesity:	Coronary artery disease, hypertension, non-insulin-dependent diabetes mellitus, breast and uterine cancer, gallstone formation, osteoarthritis, respiratory problems, obstructive sleep apnea

WATER-SOLUBLE VITAMINS

Name the water-soluble vitamins:	B-complex vitamins (B_1, B_2, B_3, B_5, B_6, B_{12}), folate, vitamin C, and biotin (Fig. 3.2)

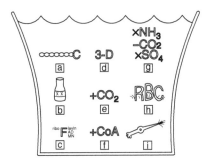

Figure 3.2 Water-soluble vitamins. (a) Vitamin C; (b) Vitamin B_1; (c) Vitamin B_2; (d) Vitamin B_3; (e) Biotin; (f) Vitamin B_5; (g) Vitamin B_6; (h) Folate; (i) Vitamin B_{12}.

What is the physiological function of vitamin B_1 (thiamine)?	Cofactor in the hexose monophosphate (HMP) shunt and in the oxidative decarboxylation of α-ketoacids (ie, pyruvate, α-ketoglutarate)
Which two clinical syndromes are associated with thiamine deficiency, particularly in alcoholics?	1. Beriberi 2. Wernicke-Korsakoff syndrome
What are the clinical features of beriberi syndrome?	Polyneuritis (*dry*), dilated cardiomyopathy (*wet*), and edema
What are the clinical features of Wernicke-Korsakoff syndrome?	Gait disturbance, diplopia (for Wernicke encephalopathy), confabulation, and memory loss (for Korsakoff psychosis)
Name two pathways that utilize folate as a cofactor:	1. Purine 2. Pyrimidine synthesis
What are the two most common causes of folate deficiency?	1. Pregnancy 2. Alcoholism
What is the danger associated with folate deficiency during pregnancy?	Neural tube defects (ie, spina bifida, anencephaly)
What characteristic features are found on a peripheral blood smear in patients with folate deficiency?	Megaloblastic (macrocytic) anemia

Name five medications that can interfere with folate utilization in the body:

1. Dilantin, phenytoin (anticonvulsants)
2. Methotrexate (cancer and rheumatoid arthritis)
3. Sulfasalazine (Crohn disease and ulcerative colitis)
4. Trimethoprim (antibacterial)
5. Pyrimethamine (antimalarial)

What is the mechanism of action of sulfonamide antibiotics?

Sulfonamide antibiotics mimic p-aminobenzoic acid (PABA), a folate precursor, competing with the physiological precursor for the enzyme involved in the next step of folate synthesis.

Which vitamin is a cofactor in both gluconeogenesis and fatty acid synthesis?

Biotin

Name three reactions that use biotin as a cofactor:

1. Carboxylations of pyruvate to oxaloacetate
2. Acetyl coenzyme A (CoA) to malonyl CoA
3. Propionyl CoA to methylmalonyl CoA

Name some clinical features of biotin deficiency:

Hair loss, bowel inflammation, muscle pain, dermatitis

What vitamin deficiency is associated with the excessive consumption of raw eggs?

Biotin deficiency (avidin strongly binds biotin)

Which coenzymes are derived from vitamin B_3 (niacin) and in which types of reactions are these coenzymes used?

Nicotinamide adenine dinucleotide, NAD(H) and nicotine adenine dinucleotide phosphate, NADP(H); used in oxidation/reduction reactions

What clinical features are typically associated with pellagra?

The 4 D's:
1. Dementia
2. Dermatitis
3. Diarrhea
4. Death (if untreated)

Name three common causes of niacin deficiency:

1. Isoniazid (INH) treatment
2. Hartnup disease (autosomal dominant disease with impaired transport of neutral amino acids in the kidneys and small intestine)
3. Malignant carcinoid syndrome

What coenzymes are derived from vitamin B_2 (riboflavin)?	Flavin adenine dinucleotide, FAD^+ ($FADH_2$)
What clinical features are commonly associated with vitamin B_2 (riboflavin) deficiency?	Corneal neovascularization, cheliosis/stomatitis, and magenta-colored tongue
Vitamin B_5 (pantothenate) is used as a precursor in which coenzyme?	CoA
What are the common clinical features of pantothenate deficiency?	Adrenal insufficiency, enteritis, dermatitis, and hair loss
How is vitamin B_6 (pyridoxine) activated in the body?	Pyridoxine is converted into pyridoxal phosphate
Which enzymes make use of this active form?	Aminotransferase (aspartate aminotransferase, alanine aminotransferase [AST, ALT]), decarboxylation, and transsulfuration enzymes
What are the clinical features of vitamin B_6 deficiency?	Convulsions, hyperirritability, cheliosis/stomatitis, and sideroblastic anemia
Name two drugs that can induce a deficiency of vitamin B_6:	1. INH 2. Oral contraceptives
High doses of vitamin B_6 may be used to treat what disease?	Homocystinuria
How is vitamin B_{12} (cobalamin) synthesized, absorbed, and stored by the body?	Synthesized by microorganisms in the small intestine, absorbed by an intrinsic factor–mediated mechanism in the terminal ileum, and stored in the liver
What are some dietary sources of vitamin B_{12}?	Animal products including eggs, meat, and dairy
What dietary lifestyle is associated with vitamin B_{12} deficiency?	Veganism—vegans exclude all animal products from their diets
Name two enzymes that require B_{12} as a cofactor:	1. Methionine synthase (or 5-methyltetrahydrofolate homocysteine methyltransferase) in homocysteine methylation 2. Methylmalonyl CoA mutase in methylmalonyl CoA processing

List some causes of vitamin B$_{12}$ deficiency:	Celiac disease, enteritis, *Diphyllobothrium latum* infection (tapeworm), alcoholism, Crohn disease, strict vegan/vegetarian diet, gastrectomy, terminal ileum resection, or bacterial overgrowth in the small intestine (deficiency is usually due to malabsorption, not insufficient dietary supply)
What is the pathophysiology of pernicious anemia?	Deficiency of intrinsic factor in gastric secretions (may be caused by antiparietal cell antibodies) resulting in decreased vitamin B$_{12}$ absorption
What are two common clinical consequences of vitamin B$_{12}$ deficiency?	1. Megaloblastic (macrocytic) anemia 2. Progressive peripheral neuropathy
What laboratory test can help uncover the cause of low serum B$_{12}$ levels?	Schilling test: in part I, radio-labeled vitamin B$_{12}$ is given by mouth and a large dose of unlabeled vitamin B$_{12}$ is given by intramuscular (IM) injection to saturate cellular uptake mechanisms. If the amount of label appearing in the urine over 24 hours is low, due to decreased gut absorption, then part II tests whether any identified malabsorption of vitamin B$_{12}$ is due to a lack of intrinsic factor by giving exogenous intrinsic factor in a repeat test. (Fig. 3.3)

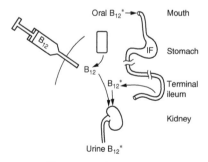

Figure 3.3 Schilling test.

Name three medications that can decrease levels of vitamin B$_{12}$:

1. Metformin
2. Phenytoin
3. Methotrexate

Name three physiological roles of vitamin C (ascorbic acid):

1. Hydroxylation of proline and lysine residues in collagen synthesis
2. Facilitation of iron absorption in the gastrointestinal (GI) tract by maintaining iron in the more absorbable reduced state (Fe^{2+})
3. Cofactor utilized in the conversion of dopamine to norepinephrine

What are some dietary sources of vitamin C?

Citrus fruits and green vegetables (deficiency is associated with a diet low in these sources)

Name some clinical features of vitamin C deficiency (scurvy):

Poor wound healing, easy bruising, bleeding gums, glossitis, and increased bleeding time

FAT-SOLUBLE VITAMINS

Which vitamins are dependent upon lipid emulsification for absorption?

The fat-soluble vitamins (vitamins A, D, E, and K) (Fig. 3.4)

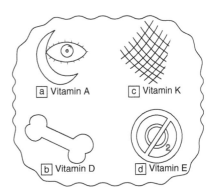

Figure 3.4 Fat-soluble vitamins.

Name some conditions that can cause a deficiency of fat-soluble vitamins:

Tropical or celiac sprue, cystic fibrosis, chronic pancreatitis, lactose intolerance, extended antibiotic use, and excessive mineral oil intake

Do water-soluble or fat-soluble vitamins more often cause toxicity? Why?

Fat-soluble vitamins, because they accumulate in fat

How is vitamin D synthesized by the body?

7-Dehydrocholesterol, derived from cholesterol, is synthesized in the liver; UV rays on sun-exposed skin convert 7-dehydrocholesterol into cholecalciferol (the storage form of vitamin D)

Name one dietary source of vitamin D:

Commercial dairy products (contain UV-irradiated ergocalciferol derived from yeast)

How is vitamin D (cholecalciferol) converted to its active form?

Stored cholecalciferol in the liver is hydroxylated to 25-hydroxycholecalciferol by liver enzymes; when serum calcium is low, the parathyroid gland releases parathyroid hormone (PTH) which activates 1α-hydroxylase to convert 25-hydroxycholecalciferol to 1,25-dihydroxycholecalciferol (1,25-DHCC) in a second hydroxylation that occurs in the kidneys.

What is the role of active 1,25-DHCC in calcium homeostasis?

Increases intestinal calcium uptake by acting as a lipid-soluble hormone on duodenal epithelia and increases calcium reabsorption in the kidney; 1,25-DHCC also acts in conjunction with PTH to mobilize the calcium stores found in bone.

Name three medical conditions that decrease vitamin D activity:

1. End-stage renal disease (resulting in decreased 1α-hydroxylase activity in proximal tubule cells)
2. Fanconi syndrome (a defect in proximal renal tubule cells)
3. Genetic deficiency of the 1α-hydroxylase enzyme (vitamin D-resistant rickets)

What condition is the result of vitamin D deficiency during childhood?	Rickets
What are the clinical features of this disorder?	Skeletal abnormalities (bowing deformities of the legs and other developing bones), osteomalacia (bone softening due to vitamin D deficiency following epiphyseal closure)
What is the difference between rickets and osteomalacia?	Rickets and osteomalacia generally occur together as long as the epiphyseal growth plates are open (childhood); only osteomalacia occurs after epiphyseal closure (adulthood).
What symptoms do you expect to find with hypercalcemia?	Stupor, change in mental status, nausea/vomiting, flank pain, and polyuria
Which drugs can cause hypercalcemia when taken along with vitamin D and/or related compounds?	Thiazide diuretics (decrease calcium excretion in the kidney)
What is the role of vitamin A in epithelial cells?	Growth, differentiation, and maintenance

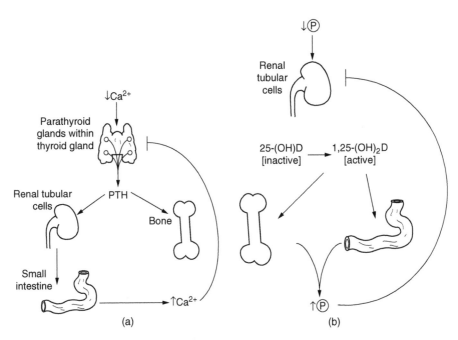

Figure 3.5 Calcium homeostasis. (a) PTH regulation; (b) Vitamin D regulation.

What is the role of vitamin A in visual pigments?

Converted into *cis*-retinal and acts as a cofactor for opsin (protein that synthesizes rhodopsin which is utilized in rods for night vision)

What is the role of vitamin A in the immune system?

Maintenance of mucous membranes, plays a role in leukocyte function

Name some potential signs of vitamin A deficiency:

Night blindness, dry eyes, dry skin, bronchitis, pneumonia, and impaired immune response

Name some consequences of vitamin A excess:

Arthralgia, headaches, fatigue, skin changes, sore throat, and hair loss

Name a common clinical use of retinol:

Retinoic acid is often used in the treatment of acne

What is the major physiological role of vitamin E?

Antioxidant; also counteracts atherosclerotic changes and coronary artery disease by preventing the oxidation of low-density lipoprotein (LDL) particles

List the dietary sources of vitamin E:

Green leafy vegetables and seed grains

What is the most significant feature of vitamin E deficiency?

Erythrocytes will demonstrate increased osmotic and peroxidative fragility, causing hemolytic anemia

Name a test which can be used to help diagnose vitamin E deficiency:

Osmotic fragility test (also used to test for hereditary spherocytosis)

What is the most significant source of vitamin K in a healthy person?

Synthesis by intestinal bacteria

What is the physiological role of vitamin K?

γ-Carboxylation of the coagulation factors II, VII, IX, X, and proteins C and S

Name some causes of vitamin K deficiency:

Fat malabsorption syndromes, long-term antibiotic therapy (removes intestinal flora), breast-feeding infants (lack intestinal flora), anticonvulsant drugs (ie, phenyldantoins)

What are the clinical features of vitamin K deficiency?	Mild vitamin K deficiency will prolong prothrombin time (PT), but partial thromboplastin time (PTT) will be normal; in neonates, vitamin K deficiency will cause hemorrhage with increased PT and activated partial thromboplastin time (aPTT), but normal bleeding time; severe vitamin K deficiency will prolong PT and PTT.
What are the clinical features of vitamin K toxicity in infants?	Anemia, hyperbilirubinemia, kernicterus
Name a drug which antagonizes the physiological activity of vitamin K:	Warfarin (Coumadin)
How does Warfarin work as an anticoagulant?	Inhibits epoxide reductase (the enzyme which recycles vitamin K to its reduced form) thereby diminishing vitamin K stores and inhibiting production of active coagulation factors

MINERALS

What bodily functions require calcium?	Formation of bones and teeth, nerve and muscle function, blood clotting
What are the dietary sources of calcium?	Dairy products, leafy green vegetables, fortified foods
What are the signs and symptoms of calcium deficiency?	Paresthesias, Trosseau and Chvostek signs, increased neuromuscular excitability, muscle cramps, bone fractures, osteomalacia
What are Trousseau and Chvostek signs?	Trousseau sign is seen as a spasm of the hand when the brachial artery is manually occluded; Chvostek sign is elicited when tapping of the facial nerve causes twitching of the nose or lips.
Iodine is required for the formation of what biologically important hormone?	Thyroid hormone
What are the dietary sources of iodine?	Seafood (ie, shellfish), iodized salt

What are the signs and symptoms of iodine deficiency?	Goiter, cretinism
Iron is utilized by what biologically important molecules?	Hemoglobin, myoglobin, cytochromes, oxidases, oxygenases
What foods have high iron content?	Liver, heart, wheat germ, egg yolks, oysters, fruits, dried beans
What conditions predispose to low iron levels?	Pregnancy, donating blood, old age, Crohn or Celiac disease, and adolescence
Where in the GI tract is iron absorbed?	Duodenum
Is heme iron or nonheme iron absorbed more efficiently?	Heme iron
What substances enhance iron absorption?	Vitamin C, reducing sugars, meat
What substances reduce iron absorption?	Antacids, fiber, oxalate
What are the signs and symptoms of iron deficiency?	Hypochromic microcytic anemia, fatigue, pallor, tachycardia, shortness of breath on exertion, depapillation of the tongue, pica (abnormal appetite for nonfoods, such as clay, or for nonnutritive food items, such as flour)
What is hemochromatosis?	Iron toxicity with increased deposition of iron in many organs
What is the classic triad of hemochromatosis?	Micronodular pigment cirrhosis, *bronze* diabetes, skin pigmentation
What stain is used histologically to diagnose hemochromatosis?	Prussian blue
Hemochromatosis can predispose an individual to what diseases?	Congestive heart failure, hepatocellular carcinoma
What is the treatment for hemochromatosis?	Repeated phlebotomy, IV deferoxamine
What are the physiological roles of magnesium in the body?	Binding to the active site of enzymes, complexing with adenosine triphosphate (ATP)

List the richest dietary sources of magnesium:	Dairy products, grains, nuts
Magnesium deficiency is often seen in individuals with what concurrent disorders?	Alcoholism, fat malabsorption syndromes, hypocalcemia
What are the signs and symptoms of magnesium deficiency?	Increased neuromuscular excitability, depression of PTH release (with severe hypomagnesemia) leading to hypocalcemia
Describe the roles of phosphorous (phosphate) in the body:	Structural when found in bone or DNA/RNA, buffer when found in blood, energy storage when found in ATP, compartmentalizing when found in plasma membranes
What dietary sources are rich in phosphorous?	Seafood, nuts, grains, legumes, cheese
What is the most common cause of phosphorous deficiency?	Renal failure
What are the signs and symptoms of phosphorous deficiency?	Defective bone mineralization with retarded growth, skeletal deformities, bone pain, decreased 2,3-bisphosphoglycerate leading to tissue hypoxia
Zinc is required for the function of what physiologically important molecules?	Metalloenzymes
What dietary sources are rich in zinc?	Meat, eggs, seafood, whole grains
What are the signs and symptoms of zinc deficiency?	Growth retardation, hypogonadism, impaired taste and smell, poor appetite, reduced immune function, mental lethargy, dry and scaly skin
What are the signs and symptoms of zinc toxicity?	Vomiting, diarrhea; neurological damage upon inhalation of zinc oxide fumes

SUMMARY CHARTS

Chart 3.1 Water-Soluble Vitamins

Vitamin	Pathways	Causes of Deficiency	Consequences of Deficiency
Thiamine B_1	- HMP shunt pathway - Oxidative decarboxylation of α-ketoacids	- Chronic alcoholism	- Beriberi and Wernicke-Korsakoff syndromes - Nervous system and heart show particular sensitivity to B_1 deficiency
Riboflavin B_2 (FMN, FAD^+)	- Oxidation/reduction reactions	- Insufficient dietary intake	- Corneal neovascularization, cheliosis/stomatitis, and magenta-colored tongue
Niacin B_3 (NAD, $NADP$)	- Oxidation/reduction reactions	- Isoniazid (INH) treatment, Hartnup disease, and malignant carcinoid syndrome	- Pellagra (remember the 3 D's: dementia, dermatitis, diarrhea)
Pantothenate B_5	- Used to make CoA	- Very rare	- Adrenal insufficiency, enteritis, dermatitis, and hair loss
Pyridoxine B_6 (pyridoxal phosphate)	- Aminotransferase, decarboxylation, and transsulfuration reactions	- INH and oral contraceptives	- Convulsions, hyperirritability, cheliosis/stomatitis, and sideroblastic anemia

Biotin B$_7$	- Carboxylation reactions (ie, pyruvate to oxaloacetate, acetyl CoA to malonyl CoA, and propionyl CoA to methylmalonyl CoA	- Excessive consumption of raw eggs (avidin strongly binds biotin)	- Hair loss, bowel inflammation, muscle pain, dermatitis
Cobalamin B$_{12}$	- Cofactor for homocysteine methyltransferase and methylmalonyl CoA mutase	- Malabsorption problems (ie, Celiac disease, tapeworm infection, alcoholism, Crohn disease, enteritis, etc.)	- Megaloblastic anemia and progressive peripheral neuropathy
Folate B$_9$	- Nucleotide synthesis	- Pregnancy increases the need for folate - Alcoholism	- Neural tube defects (ie, spina bifida, anencephaly)
Ascorbic acid C	- Hydroxylation of proline & lysine residues in collagen synthesis - Facilitation of iron absorption in GI tract - Cofactor in the dopamine to norepinephrine conversion	- Diet low in citrus fruits and green vegetables	- Scurvy (symptoms: poor wound healing, easy bruising, bleeding gums, glossitis, and increased bleeding time

Chart 3.2 Fat-Soluble Vitamins

Vitamin	Roles	Signs/Symptoms of Deficiency	Signs/Symptoms of Toxicity
Vitamin A	- Growth, differentiation, and maintenance of epithelial cells - Night vision - Immune system	- Night blindness, dry eyes, dry skin, bronchitis, pneumonia, and impaired immune response	- Arthralgia, headaches, fatigue, skin changes, sore throat, and hair loss
Vitamin D (cholecalciferol, active form = 1,25 dihydroxycholecalciferol)	- Calcium homeostasis (promotes intestinal calcium uptake & increases calcium reabsorption in the kidney; also acts with PTH to mobilize calcium bone stores)	- Rickets (during childhood), skeletal abnormalities, osteomalacia	- Hypercalcemia
Vitamin E (tocopherols & tocotrienols)	- Antioxidant	- Hemolytic anemia	- Very rare
Vitamin K	- γ-Carboxylation of the coagulation factors II (prothrombin), VII, IX, X, and proteins C and S	- Long prothrombin time - Hemorrhagic disease in newborns - Excess bleeding	- Anemia, hyperbilirubinemia, kernicterus in infants

Chart 3.3 Minerals

Mineral	Roles	Signs/Symptoms of Deficiency	Signs/Symptoms of Toxicity
Calcium	- Formation of bones and teeth, nerve and muscle function, blood clotting	- Paresthesias - Increased neuromuscular excitability - Muscle cramps - Bone fractures - Osteomalacia	
Iodine	- Required for formation of thyroid hormone	- Goiter, cretinism	
Iron	- Utilized by hemoglobin, myoglobin, cytochromes, oxidases, and oxygenases	- Hypochromic microcytic anemia - Fatigue - Shortness of breath on exertion - Pallor - Tachycardia - Depapillation of the tongue - Pica	- Hemochromatosis
Magnesium	- Binding to the active site of enzymes, complexing with ATP	- Increased neuromuscular excitability - Depression of PTH release leading to hypocalcemia	

(Continued)

Chart 3.3 Minerals (Continued)

Mineral	Roles	Signs/Symptoms of Deficiency	Signs/Symptoms of Toxicity
Phosphorus	- DNA/RNA, bone mineralization, buffer when found in blood, energy storage when found in ATP, compartmentalizing when found in plasma membranes	- Defective bone mineralization with retarded growth - Skeletal deformities - Bone pain - Decreased 2,3-bisphosphoglycerate leading to tissue hypoxia	
Zinc	- Required for the function of metalloenzyme	- Growth retardation - Hypogonadism - Impaired taste and smell - Poor appetite - Reduced immune function - Mental lethargy - Dry and scaly skin	- Vomiting - Diarrhea - Neurological damage upon inhalation of zinc oxide fumes

CLINICAL VIGNETTES—MAKE THE DIAGNOSIS

A 43-year-old African American man arrives at the emergency room with his son confused, short of breath, and smelling grossly of alcohol. Additionally, his gait showcases a significant foot drop. The son reports that the father's diet consists mainly of snacks and canned foods when he is not drinking alcohol. The son also reports the father's tendency to make up elaborate stories when discussing himself. Physical examination reveals significant hepatomegaly and decreased deep tendon reflexes. Chest x-ray (CXR) shows cardiomegaly with basal lung congestion.

What vitamin may be deficient in this patient?	Thiamine, vitamin B_1
The cardiac failure and polyneuropathy encountered are manifestations of which disease?	Wet beriberi and dry beriberi, respectively
In what order should deficient nutrients be administered to this patient?	Thiamine first, followed by glucose, folate, and other vitamins

A 39-year-old alcoholic man reports the insidious onset of "not feeling like himself" and "forgetting things." His review of systems is significant for the coinciding appearance of chronic diarrhea. He reports that his diet consists of many wheat and corn products. The physical examination is significant for an erythematous, nonpruitic, hyperpigmented, scaling rash of the face, neck, and dorsum of the hands.

The constellation of signs and symptoms suggests what disease?	Pellagra
This disease is a result of what vitamin deficiency?	Niacin, vitamin B_3
What is the treatment of choice?	Oral nicotinamide

A 17-month-old boy is brought to his pediatrician by his parents for concern regarding abnormal outward bowing of his lower extremities. In addition, the child exhibits lineal chest depression along the diaphragm, and enlargement of the costochondral junctions. The child's diet is deficient in dairy products and he spends a lot of his time playing indoors.

This child is suffering from what disease?	Rickets
What would you expect this child's serum calcium, serum phosphorous, and alkaline phosphatase levels to be?	Normal/slightly low, decreased, and increased, respectively
What are potential treatments for this child's condition?	Increase egg and dairy product intake, increase exposure to sunlight

A 42-year-old man who recently emigrated from Sudan sustains a femoral neck fracture when he accidentally slips in the shower. He reports that the fall was not too severe. He has been suffering from recurrent lower back pain and leg weakness.

A lumbar-sacral spine x-ray would most likely show what findings?	Collapse of lumbar vertebrae and generalized osteopenia
This man is suffering from what condition?	Osteomalacia
What are potential etiologies for this man's condition?	Lack of sunlight exposure, intestinal malabsorption, renal insufficiency, target organ resistance

A 9-month-old girl with poor perinatal care is brought to the pediatrician because of listlessness, pallor, and anorexia. The child's gums bleed easily and she has petechiae over her nasal and oral mucosa. Her coagulation tests reveal a prolonged bleeding time.

This patient likely suffers from what disease?	Scurvy
What nutrient could this child be lacking in her diet?	Ascorbic acid, vitamin C
What role does this nutrient play in collagen formation?	Ascorbic acid hydroxylates praline and lysine
What other populations are at risk for developing this condition?	Smokers, oncologic patients, alcoholics, elderly

A 33-year-old Mexican immigrant is seen in the emergency department after crashing his taxi cab late that night. He claims to not have suffered any major injuries but does mention that his visual acuity at night has decreased significantly since he arrived in the United States 3 years ago.

What physical signs can you look for to confirm this man's condition?	Conjunctival xerosis, Bitot spots
This man's diet is deficient in which vitamin?	Retinol, vitamin A
How is this vitamin involved in visual acuity?	Retinol is used for the synthesis of rhodopsin in the retina

An elderly-appearing alcoholic man with a history of seizures comes to the emergency room with hematemesis, hemarthrosis of his left knee, bleeding gums, and generalized weakness. The man reports that he takes phenytoin as an antiepileptic medication. Physical examination reveals a thin and malnourished man with subcutaneous ecchymosis in his arms and legs. Coagulation tests show a prolonged prothrombin time (PT) and partial thromboplastin time (PTT).

This man's presentation is consistent with deficiency of which vitamin?	Vitamin K
Which coagulation factors are dependent upon this vitamin for their activity?	Factors II, VII, IX, and X (via γ-carboxylation)

| What other situations can predispose to this type of bleeding? | Broad-spectrum antibiotic use, malabsorption, dietary vitamin K deficiency |

A 27-year-old woman presents to her physician with weakness, easy fatigability, nausea, and diarrhea but no neurological signs. She reports that she does not regularly eat green leafy vegetables. A complete blood count (CBC) shows hypersegmented polymorphonuclear neutrophils (PMNs) and a megaloblastic anemia.

This woman's diet is likely deficient in which nutrient?	Folate
What is the importance of this nutrient?	Folate is used in the synthesis of DNA and RNA; it also acts as a coenzyme for one-carbon transfer and is involved in methylation reactions
If this woman becomes pregnant, her fetus would be at risk for what congenital abnormalities?	Neural tube defects

A 40-year-old Caucasian gentleman is admitted with a hematocrit of 14. He is noted to have a lemon-yellow waxy pallor and a painful swollen beefy tongue. Neurological examination revealed paresthesias, weakness, and an unsteady gait. A peripheral blood smear shows a macrocytic anemia. Mean corpuscular volume (MCV) and mean cell hemoglobin (MCH) are increased with normal mean corpuscular hemoglobin volume (MCHV).

This patient is suffering from what?	Pernicious anemia
This disease is characterized by a lack of what in gastric secretions?	Intrinsic factor
The above is used for the absorption of what?	Vitamin B_{12}

A recently adopted 2-year-old child of African descent is brought to the pediatric clinic by his Caucasian foster parents. They are concerned about his general appearance. The child appears generally cachectic with a large abdomen, depigmented skin and hair, and generalized pitting edema. The child is below the fifth percentile in height and weight.

This child is suffering from what form of malnutrition?	Kwashiorkor (protein deprivation with normal total caloric intake)
What is the cause of the generalized pitting edema?	Hypoalbuminemia
What changes would be expected to be seen in the liver?	Fatty infiltration

A 25-year-old girl complains of easy fatigability and weakness of approximately 3 years' duration. Physical examination reveals pallor, tachycardia, and cheliosis. She reports that her menses have been "heavy" as long as she can recall.

This patient is likely suffering from what type of anemia?	Iron-deficiency anemia
What would microscopic examination with Prussian blue staining of her bone marrow likely show?	Erythroid hyperplasia with decreased bone marrow iron stores
What are possible treatments of this disorder?	Control menstrual blood loss, supplemental iron

A 3-year-old boy is brought to the pediatric clinic by his mother because of developmental delay. She also reports that the boy has exhibited behavior problems, such as biting his lip excessively and hitting himself. Physical examination is notable for scarring on the lips and swelling on the toes and fingers. Laboratory tests reveal hyperuricemia.

What disease is this child suffering from?	Lesch-Nyhan syndrome
Deficiency of what enzyme is the cause of this disorder?	HGPRT (hypoxanthine phosphoribosyltransferase)
What pathway is the enzyme involved in?	Purine salvage pathway
What is the inheritance pattern of this disease?	X-linked recessive

CHAPTER 4

Molecular Biology

OVERVIEW

What constitutes a nucleoside?
A nitrogenous base linked to a pentose monosaccharide

What constitutes a nucleotide?
A nitrogenous base, a pentose monosaccharide, and either 1, 2, or 3 phosphate groups (basically a phosphorylated nucleoside)

Which constituent(s) carries genetic information?
The nitrogenous base

Which constituent(s) maintains the backbone of DNA?
The sugar and phosphate groups

Name the two families of nitrogenous bases:
1. Purines
2. Pyrimidines

What is the difference between a purine and a pyrimidine?
Purines contain two rings whereas pyrimidines contain one ring.

Name the two bases in the purine class:
1. Adenine
2. Guanine (remember **PURE** As **G**old)

Name the three bases in the pyrimidine class:
1. Cytosine
2. Uracil
3. Thymine (remember **PYR**AMIDS are **CUT**)

How do uracil and thymine differ in structure?
Thymine contains a methyl group (THYmine contains meTHYl)

What bases can be found in DNA?
Adenine, guanine, thymine, and cytosine

What bases can be found in RNA?
Adenine, guanine, uracil, and cytosine

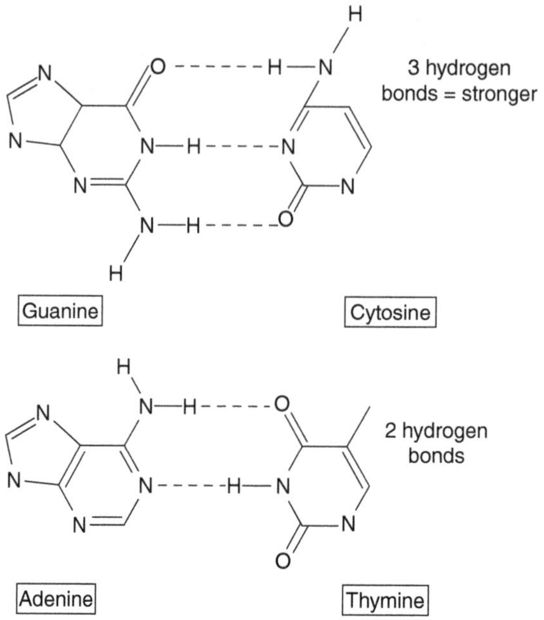

Figure 4.1   Nitrogenous bases. (a) Adenine; (b) Guanine; (c) Thymine; (d) Cytosine.

Describe the polarity of the DNA chain: One end of the chain has a 5'-OH group and the other a 3'-OH group; therefore, the base sequence is written in the 5' to 3' direction.

What bases pair with each other in the complementary strands of DNA? Adenine pairs with thymine (held together by two hydrogen bonds), guanine pairs with cytosine (held together by three hydrogen bonds).

Which base pair has a stronger bond? The guanine-cytosine base pair, because it is held together by three hydrogen bonds.

List the important features of the Watson and Crick model of DNA: Double helix, with the sugar-phosphate chains running in opposite directions.

Base pairs are on the inside of the double helix.

Helical structure repeats after 10 residues on each chain; helix turns 360° every 10 residues.

Chains are held together via hydrogen bonding between bases.

Genetic information is carried in the precise sequence of bases.

What are some key differences between DNA and RNA?	DNA is usually double-stranded (in a double-helix format), RNA is usually single-stranded; the sugar in DNA is deoxyribose, the sugar in RNA is ribose; DNA utilizes adenine, thymine, guanine, and cytosine as bases, RNA utilizes adenine, uracil, guanine, and cytosine as bases.
What are the two stages of making proteins?	1. Transcription (DNA to RNA) 2. Translation (RNA to protein)
In what direction is DNA replicated?	5′ to 3′
What are plasmids?	Small, circular, extrachromosomal DNA molecules that may or may not be synchronized with chromosomal division
What type of information can be carried by a plasmid?	The genes for inactivation of specific antibiotics, production of toxins, and/or breakdown of natural products
What is the importance of plasmids in pharmacology?	Plasmids may be used by microorganisms to confer resistance to a specific antibiotic.

LABORATORY TECHNIQUES

What separation methods can be used to yield a purified protein based upon:	
Size?	Gel filtration, preparative gel electrophoresis
Ionic charge?	Gel electrophoresis, ion-exchange chromatography
Binding to ligands or antibodies?	Affinity chromatography
List chemicals/enzymes used to selectively cleave proteins:	Trypsin, chymotrypsin, 2-nitro-5-thiocyanobenzene, cyanogens bromide
Enzyme-linked immunosorbent assay (ELISA) tests what type of reactivity?	Antigen-antibody reactivity (an antibody test to an exposed antigen)
When is ELISA used?	In the serodiagnosis of specific infectious diseases [ie, human immunodeficiency virus (HIV)] to determine whether a particular antibody is present in a patient's blood sample

What methods are used to determine the three-dimensional structure of a protein?	X-ray crystallography, nuclear magnetic resonance
Describe the action of a restriction endonuclease:	Recognizes specific base sequences (typically palindromic sequences) in double-helical DNA and cleaves both strands of the duplex at specific sites
What is a palindromic sequence?	DNA sequence that is the same when read either 5′ to 3′ or 3′ to 5′. DNA is double stranded, so base pairs must be read not just the bases on one strand to see if a sequence is palindromic.

(a): Example of palindromic sequence
    5′–G G A T C C–3′
    3′–C C T A G G–5′

(b): Example of a restriction site recognized by a restriction enzyme
    5′–G⌊G A T C C–3′
    3′–C C T A G⌉G–5′

Figure 4.2   (a) Example of a palindromic sequence; (b) Example of a restriction site recognized by a restriction enzyme.

How is the action of a restriction endonuclease of benefit in gel electrophoresis?	Small differences in base sequence between related DNA molecules result in restriction fragments of different sizes upon exposure to a restriction endonuclease; gel electrophoresis is then used to separate these fragments based on molecular size.
What is the relationship between the thickness of a band on an electrophoresis gel and the abundance of restriction fragment?	Thickness of the band is directly proportional to the abundance of restriction fragment.
What genetic material is probed in Southern blotting?	DNA
What genetic material is probed in Northern blotting?	RNA

What is a Western blot?

Separation of proteins via electrophoresis, followed by identification by specific complexing with antibodies that are tagged with a radio-labeled second protein

What is the Sanger dideoxynucleotide method used to determine?

The sequence of bases in DNA fragments

List the steps of the Sanger dideoxynucleotide method:

Denature the DNA fragment into single strands and divide them into four samples.

Add the following to each sample: an oligonucleotide primer, a large excess of all four deoxynucleoside triphosphates [deoxyadenosine triphosphate, deoxyguanosine triphosphate, deoxycytidine triphosphate, and deoxythymidine triphosphate (dATP, dGTP, dCTP, and dTTP)], DNA polymerase, and a small amount of a dideoxynucleotide triphosphate (ddNTP) analogous to one of the four DNA molecules.

To enable detection of the DNA fragments, label the primer at the 5' end or include a labeled deoxynucleotide triphosphate (dNTP) in the reaction mixture (can also use a color-specific fluorophore-labeled dNTP).

The ddNTP stops replication when it is incorporated into the growing chain because it has no 3' OH.

Subject the reaction products to gel electrophoresis and autoradiography, and read the sequence from the band patterns (see Fig. 4.3).

Describe the functionality of polymerase chain reaction (PCR):

Used to synthesize many copies of a desired fragment of DNA

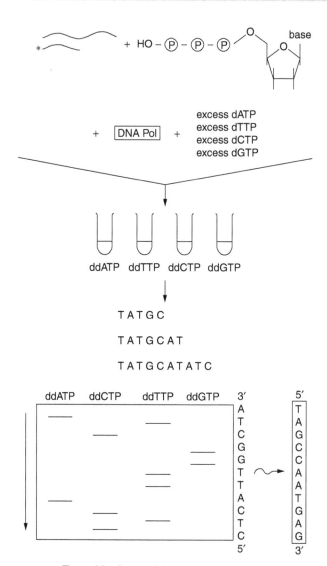

Figure 4.3   Sanger dideoxynucleotide method.

Describe the steps in PCR:	DNA is denatured, or "melted," by heating to generate two separate strands.
	During cooling, excess primers anneal to a specific sequence on each strand to be amplified.
	Heat-stable DNA polymerase replicates the DNA sequence following each primer.
	All the above steps are repeated many times. (Fig.4.4)

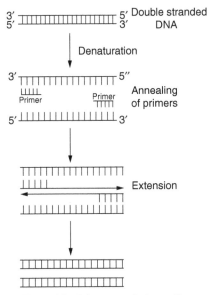

Figure 4.4   Polymerase chain reaction.

PCR can be used to detect what level of genetic mutation?	Single base pair mutation
What are the uses of DNA/protein cloning?	To amplify and obtain a large quantity for study
What is a vector?	A vehicle used to transfer genetic material to a target cell
What vectors are used to clone DNA?	Bacteriophage, plasmid, yeast artificial chromosome (YAC), bacterial artificial chromosome (BAC)
Briefly describe the microarray technique:	Nucleic acid sequences from thousands of different genes are fixed onto a surface (ie, glass slide, silicon chip). DNA or RNA probes are hybridized to the array, and the amount of complementary binding is quantified.
What are the uses of microarrays?	Gene expression profiling, detection of single nucleotide polymorphisms (SNPs)

Describe the steps involved in cloning DNA and/or protein:

(a) Cleave the DNA that is to be cloned and the DNA of the vector with the same restriction endonuclease so as to create "sticky ends."

(b) Attach the foreign DNA to the vector by treatment with DNA ligase, thus producing recombinant DNA.

(c) Transform bacterial cells by incubating them with the vector containing the recombinant DNA.

(d) Plate the transformed bacterial cells to produce individual colonies.

Identify and select the colonies containing the recombinant DNA using a probe; isolate and culture those colonies.

(e) Isolate and characterize the recombinant DNA, or protein expressed from the recombinant DNA, from the bacterial cells. (Fig. 4.5)

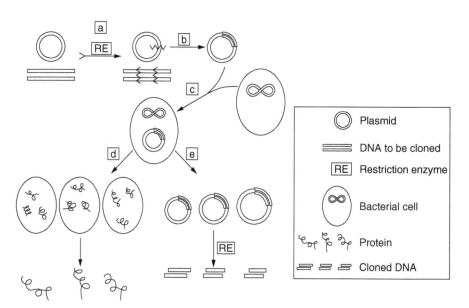

Figure 4.5   Cloning DNA and protein.

| What is a single nucleotide polymorphism? | A DNA sequence variation occurring when a single nucleotide differs between members of a species |

CELL CYCLE

| What are the phases of the cell cycle? | G_1, S, G_2, and M (G = gap, S = synthesis, M = mitosis) (Fig. 4.6) |

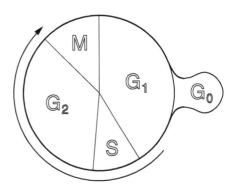

Figure 4.6 Cell cycle.

| What happens in the G_1 phase? | The cell prepares to initiate DNA synthesis; chromosomes begin to unfold and form euchromatin. |

| The rate of cell division is inversely proportional to the length of what phase of the cell cycle? | G_1 phase |

| Resting or differentiated cells are considered to be in what stage of the cell cycle? | G_0 of the G_1 phase |

| What roles do the retinoblastoma tumor suppressor protein (pRb) and E2F play in the cell cycle? | E2F is a positive transcription factor that allows the cell to progress into the S phase of the cell cycle. pRb is a protein that, if not phosphorylated, inactivates E2F by binding to it. When pRb is phosphorylated, it releases E2F and the cell is allowed to progress into the S phase of the cell cycle. |

| Mutations in pRb are associated with what types of malignancies? | Retinoblastomas, osteosarcomas |

What happens in the S phase?

Replication of DNA (DNA doubles in a semiconservative manner)

What happens in the G_2 phase?

The cell synthesizes the RNA and proteins that are required for mitosis to proceed; chromatin recondenses to form heterochromatin.

Name the four stages of mitosis:

1. Prophase
2. Metaphase
3. Anaphase
4. Telophase

Methotrexate acts upon which phase of the cell cycle?

S phase

Describe methotrexate's mechanism of action:

It is a folic acid analog, inhibiting dihydrofolate reductase, which results in decreased deoxythymidine monophosphate (dTMP) along with decreased DNA and protein synthesis.

What is methotrexate used to treat?

Rheumatoid arthritis, psoriasis, lymphomas, leukemias, sarcomas, and ectopic pregnancy

What is a common side effect associated with methotrexate use?

Myelosuppression

Is this toxicity reversible?

Yes, it can be reversed with leucovorin (folinic acid)—termed leucovorin rescue.

5-Fluorouracil (5-FU) acts upon which phase of the cell cycle?

S phase

What is the mechanism of action of 5-FU?

It is a pyrimidine analog which is converted to 5F-dUMP. 5F-dUMP binds to thymidylate synthase and this binding is stabilized by N^5,N^{10}-methylene THF binding. The combination inhibits thymidylate synthase, which decreases dTMP along with DNA and protein synthesis.

What is 5-FU used to treat?

Solid tumors, basal cell carcinomas

What is a common side effect associated with 5-FU use?

Myelosuppression (tip: "if you're forced into guessin', say myelosuppression" with the anticancer drugs)

Is this toxicity reversible?

No, this is a major difference between methotrexate and 5-FU.

6-Mercaptopurine (6-MP) acts upon which phase of the cell cycle? — S phase, indirectly

What is 6-MP's mechanism of action? — 6-MP blocks purine synthesis, thus decreasing DNA synthesis.

What enzyme activates 6-MP? — Hypoxanthine-guanine phosphoribo-syltransferase, HGPRTase (the same enzyme which is deficient in Lesch-Nyhan syndrome)

What is 6-MP used to treat? — Leukemia and some lymphomas

What are some common toxicities associated with 6-MP use? — Bone marrow, gastrointestinal (GI) system, and liver toxicities

Azathioprine is a derivative of what drug? — 6-MP

What is the mechanism of action of azathioprine? — Interferes with the synthesis of nucleic acids

What is azathioprine used to treat? — It is used as an immunosuppressant for transplantations and autoimmune disorders.

In what phase of the cell cycle does bleomycin work? — G_2 phase

Describe bleomycin's mechanism of action: — Intercalates DNA and induces free radicals, which create DNA strand breaks

What is bleomycin used to treat? — Lymphomas, testicular cancer

What are some toxicities of bleomycin? — Myelosuppression, pulmonary fibrosis

In what phase of the cell cycle do the vinca alkaloids (vincristine and vinblastine) work? — M phase

What is the mechanism of action of the vinca alkaloids? — Bind to tubulin and block the polymerization of microtubules, thus not allowing the mitotic spindle to form

What are the vinca alkaloids used to treat? — Lymphoma (vincristine is part of the mechlorethamine, oncovin, procarbazine, prednisone [MOPP] regimen), choriocarcinoma, Wilms tumor

What are the toxicities associated with the use of vinca alkaloids? — Paralytic ileus and neurotoxicity with vincristine, myelosuppression with vinblastine

What is the mechanism of action of paclitaxel?	Binds to tubulin and overstabilizes the mitotic spindle, which does not allow the breakdown of the spindle (therefore mitosis gets stuck in metaphase)
What is paclitaxel used to treat?	Breast and ovarian cancer
What is the main toxicity associated with paclitaxel use?	Myelosuppression

DNA AND REPLICATION

What is semiconservative replication?	The process by which DNA replicates itself; the two strands are pulled apart and each strand acts as a complement to the newly transcribed DNA (therefore each new DNA is half of the original copy)
What is the origin of replication?	The site at which DNA replication begins (in prokaryotes replication begins at a single nucleotide sequence, whereas in eukaryotes replication begins at multiple sites along the DNA helix)
What is a replication fork?	A replication fork is where the two strands of DNA are unwinding and separating, creating a "V," which is where active synthesis is taking place.
Is replication of DNA bidirectional or unidirectional?	Replication of DNA is bidirectional (meaning the replication forks move in both directions away from the origin).
What three prepriming complex proteins are required for the formation of a replication fork?	1. DNA protein; binds to specific nucleotide sequences at the origin of replication, usually where the parental strand is rich in AT base pairs, causing the DNA strands to separate 2. Single-stranded binding protein; stabilizes the single-stranded DNA 3. DNA helicase; forces the DNA strands apart
Which enzymes relieve supercoils in DNA?	Topoisomerases

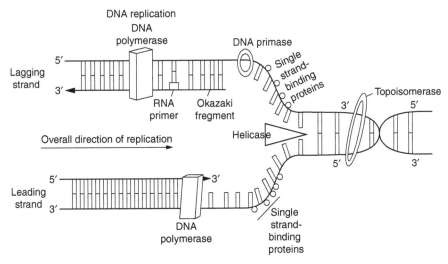

Figure 4.7   DNA replication.

Describe the action of DNA topoisomerase I:	Cuts a single strand of the DNA helix, thus relieving a supercoil
Describe the function of DNA topoisomerase II:	Creates breaks in both strands of DNA
What is the mechanism of action of etoposide?	Inhibits topoisomerase II and increases DNA degradation
What is etoposide used to treat?	Testicular cancer and oat cell carcinoma of the lung (small cell carcinoma)
What are some toxicities associated with etoposide?	Myelosuppression, alopecia, GI irritation, and peripheral neurotoxicity
What is the function of DNA polymerase in replication?	To copy the DNA templates by catalyzing the step-by-step addition of deoxyribonucleotide units to a DNA chain
What does DNA polymerase need to synthesize a chain of DNA?	All four activated precursors (ie, dATP, dGTP, dTTP, and dCTP) and Mg^{2+} Primer chain with a free 3'-OH group DNA template
In what direction does DNA polymerase read the parental nucleotide sequence?	3' to 5' direction (thus DNA replication takes place in the 5' to 3' direction)

What is the leading strand?	The strand that is copied in the direction of the advancing replication fork and is synthesized continuously
What is the lagging strand?	The strand that is copied in the opposite direction of the advancing replication fork and is synthesized rather discontinuously (thus creating small fragments of DNA called Okazaki fragments)
What links the Okazaki fragments together to form a continuous strand of DNA?	DNA ligase
What does DNA polymerase require to begin synthesis of a new strand?	RNA primer (made by primase in prokaryotes)
Which DNA polymerase is involved in the replication of DNA in prokaryotes?	DNA polymerase III
Which DNA polymerases are involved in replication of DNA in eukaryotes?	DNA polymerase α synthesizes RNA primer and replicates lagging strand. DNA polymerases β and ε are repair polymerases. DNA polymerase δ replicates the leading strand.
What is cytarabine's mechanism of action?	DNA polymerase inhibitor
What is cytarabine used to treat?	Leukemia, non-Hodgkin lymphoma
List the common toxicities associated with cytarabine use:	Thrombocytopenia, leucopenia, and megaloblastic anemia
What are some endogenous causes of DNA damage?	Base mismatch Base oxidation Base alkylation Base hydrolysis
What are some exogenous causes of DNA damage?	UV light (can cross-link adjacent cytosine and thymine bases to cause pyrimidine dimers) Ionizing radiation Extreme heat Chemicals

How are errors in base matching repaired?

DNA polymerase has a 3′ to 5′ exonuclease activity (which releases one nucleotide at a time), allowing the detection and removal of mismatched base pairs (hence "editing")—this process is in eukaryotes.

What happens if these errors cannot be repaired?

Increased incidence of mutations

What does DNA loop around to condense itself?

Positively-charged histones (H2A, H2B, H3, H4)

What structure does the aforementioned complex create?

Nucleosome beads (Fig. 4.8)

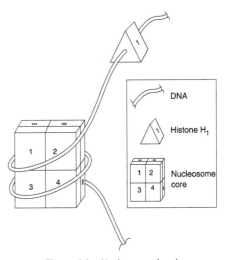

Figure 4.8 Nucleosome bead.

What protein bridges adjacent nucleosome beads to form 30-nm fiber?

H1 histones

What is chromatin?

Condensed DNA looped twice around histone "beads" or DNA plus histones

What is heterochromatin?

The more condensed, transcriptionally inactive form of chromatin

What is euchromatin?

The less condensed, transcriptionally active form of chromatin (remember "eu" means "true," so euchromatin is "truly" transcribed)

What are some of the ways in which DNA can become damaged?

Ultraviolet light exposure, extremes of pH, increased temperature, and alkylating agents

Name and describe four DNA repair mechanisms:

1. Nucleotide excision repair: an oligonucleotide of the damaged bases is excised and released by an endonuclease. DNA polymerase adds in new nucleotides. DNA ligase seals the gap.
2. Base excision repair: damaged bases are removed by glycosylases. AP endonuclease removes the AP (abasic) site and the neighboring nucleotides. The gap is filled by DNA polymerase and ligase.
3. Mismatch repair: mismatched nucleotides on the unmethylated, new strand are removed by DNA polymerase's 3' to 5' exonuclease activity. The gap is filled by DNA polymerase and ligase.
4. Nonhomologous end joining: two ends of DNA fragments are brought together.

Name four DNA repair defect syndromes:

1. Xeroderma pigmentosum (exposure to UV light)
2. Ataxia-telangiectasia (exposure to x-rays)
3. Bloom syndrome (exposure to radiation)
4. Fanconi anemia (exposure to intercalating agents)

What DNA repair mechanism is mutated in xeroderma pigmentosum?

Nucleotide excision repair

What DNA repair mechanism is mutated in hereditary nonpolyposis colorectal cancer (HNPCC)?

Mismatch repair

What is a transition mutation?

When a purine is substituted for a purine or a pyrimidine is substituted for a pyrimidine

What is a transversion mutation?

When a purine is substituted for a pyrimidine or vice versa (think transconversion)

Name some features of the genetic code:	It is unambiguous (each codon specifies only one amino acid), degenerate (more than one codon codes for a single amino acid), comma-less and nonoverlapping, and universal.
What is a silent DNA mutation?	A mutation in a codon that results in a new codon which codes for the same amino acid (often the third base in the codon is changed, ie, the wobble position)
What is a missense DNA mutation?	A mutation in a codon that results in a new codon which codes for a different amino acid
What is a nonsense DNA mutation?	A mutation that creates a stop codon and prematurely terminates the translation of a coded protein
What is a frameshift DNA mutation?	A mutation that inserts or deletes bases resulting in a misreading of the genetic code downstream of the mutation; usually resulting in a premature stop codon
What is the order of severity of the aforementioned DNA mutations?	Nonsense/frameshift mutation > missense mutation > silent mutation

RNA AND TRANSCRIPTION

What is transcription?	The conversion of DNA to RNA (transcripts of certain regions of DNA)
What are the three major types of RNA?	1. Ribosomal RNA (rRNA) 2. Messenger RNA (mRNA) 3. Transfer RNA (tRNA)
Name the different types of eukaryotic RNA polymerases:	RNA polymerase I makes rRNA, RNA polymerase II makes mRNA, and RNA polymerase III makes tRNA. (Tip: remember **R-M-T**: rRNA is rampant, mRNA is massive, tRNA is tiny. I, II, and III are numbered as their products are used in protein synthesis.)
What are the different types of prokaryotic RNA polymerases?	One RNA polymerase makes all prokaryotic RNA

What are the three phases of transcription?	1. Initiation 2. Elongation 3. Termination (initiation begins with RNA polymerase II binding to a specific region of DNA known as the promoter region—usually a stretch of about six nucleotides known as the TATA box [TATAAT], about 8-10 nucleotides upstream from the start of transcription)
In what direction does transcription take place?	5′ to 3′
Describe the elongation phase:	The DNA is unwound with negative supercoils relieved by gyrases and topoisomerase II (both of these cut two strands of DNA) and positive supercoils relieved by topoisomerase I (this cuts only one strand of DNA).
What nucleotides are required for DNA elongation to occur?	dATP, dTTP, dCTP, dGTP
How does RNA polymerase differ from DNA polymerase?	RNA polymerase does not require a primer and has no exonuclease or endonuclease activity (no proofreading).
Describe the termination phase:	A termination signal is reached, which ends transcription.
What are some antibiotics that prevent cell growth by inhibiting RNA synthesis?	Rifampin (DNA-dependent RNA polymerase inhibitor) and dactinomycin (acts by binding to the DNA template and interfering with the movement of RNA polymerase along the DNA)
Describe the eukaryotic promoter regions:	A TATA box, which is near the site of transcription, and a CAAT box, which is more distant
What are enhancers?	DNA sequences that increase the rate of transcription initiation by RNA polymerase II by binding transcription factors
Are enhancers position dependent or position independent?	Position independent (meaning that they can be located upstream, within, or downstream of the area being transcribed)

What is α-amanitin's mechanism of action?	Forms a tight complex with the RNA polymerase II, thus inhibiting mRNA synthesis and eventually protein synthesis (α-amanitin is a poison produced by the *Amanita phalloides* mushroom)
Is prokaryotic or eukaryotic mRNA posttranscriptionally modified?	Only eukaryotic mRNA; prokaryotic mRNA is identical to its primary transcript heterogeneous nuclear RNA (hnRNA)
Where in the cell is mRNA transcribed?	Nucleus of eukaryotes, cytosol of prokaryotes
Where in the cell is mRNA translated?	Cytosol
What three modifications does mRNA undergo?	1. Capping of the 5′ end 2. Addition of a poly-A tail to the 3′ end 3. Removal of introns
What is an intron?	Intervening sequence that does not code for proteins
What is an exon?	Sequence that codes for proteins; they are spliced together to form mature mRNA
What do small nuclear RNAs (snRNAs) do?	They are part of small nuclear ribonucleoproteins (snRNPs), which facilitate the splicing of exon segments.
How is systemic lupus erythematosus (SLE) associated with snRNPs?	SLE results from an autoimmune response in which the patient produces antibodies against snRNPs (this is one of the associations of SLE, not the definitive association).

TRANSLATION

What is the dogma of expression for genetic information?	DNA to RNA to protein (the first step being transcription, the second step being translation)
In what direction does DNA replication, transcription, and translation occur?	5′ to 3′
What is a codon?	Three nucleotides (bases) set in a particular order that correspond to a protein

What is the usual start codon in prokaryotes?

AUG—this codes for a methionine, which is usually formulated

Name the stop codons:

UGA, UAA, UAG (remember **U G**o **A**way, **U A**re **A**way, and **U A**re **G**one)—these codons signify to the protein to stop translating the protein

Figure 4.9   The codons.

What reads the mRNA codon and delivers the correct amino acid?

tRNA

Describe the shape of tRNA:

Cloverleaf, with an anticodon area to read the mRNA's codon (see Fig. 4.10)

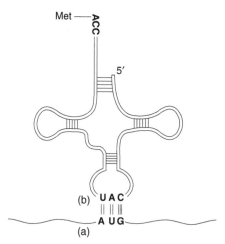

Figure 4.10   Structure of tRNA. (a) Codon; (b) Anticodon.

Where on the tRNA is the attachment for each specific amino acid?	The 3′ end
What does aminoacyl-tRNA synthetase do?	Catalyzes a two-step reaction that results in the covalent attachment of an amino acid to the 3′ end of the corresponding tRNA (this enzyme requires ATP); basically it recognizes the correct tRNA to the corresponding amino acid
Describe the structure of ribosomes in prokaryotes and eukaryotes:	Prokaryotic ribosomes have two subunits, one being the 50S subunit and the other, the 30S subunit; eukaryotic ribosomes have two subunits also, one being the 60S subunit and the other, the 40S subunit.
What are the binding sites on the ribosome for the tRNA molecule called?	The A and P sites
Explain how the A and P sites work:	They cover two neighboring codons; the A site binds an incoming aminoacyl tRNA and the P site binds the next aminoacyl tRNA to be added.
In what direction are proteins constructed?	N terminus to C terminus
Where in the cell are ribosomes located?	In eukaryotes, there are free ribosomes in the cytosol (responsible for making proteins that will stay within the cell) and ribosomes located on the rough endoplasmic reticulum (responsible for making proteins that will be secreted extracellularly, incorporated into the plasma membrane, or included within lysosomes).
How many high-energy phosphate bonds are needed to add one amino acid to a growing peptide chain?	Four bonds (two ATP for the aminoacyl-tRNA synthetase reaction + one GTP for binding the aminoacyl tRNA to the A site + one GTP for the translocation step)
Describe the "wobble" theory:	The hypothesis that if the first two nucleotides in a codon are similar and the third nucleotide is different, often the same amino acid will be coded
What is meant by polycistronic mRNA?	When a single mRNA has many coding regions, each with its own initiation site

What is meant by monocistronic mRNA?
When an mRNA codes for only one specific polypeptide

Is prokaryotic mRNA polycistronic or monocistronic?
Polycistronic (remember prokaryotes, polycistronic)

Is eukaryotic mRNA polycistronic or monocistronic?
Monocistronic

What are the three stages of translation?
1. Initiation
2. Elongation
3. Termination

What is the enzyme that catalyzes the formation of peptide bonds?
Peptidyl transferase

What is a polysome?
A complex consisting of one mRNA and multiple ribosomes translating the message

How are proteins modified following translation?
Proteins may be phosphorylated, glycosylated, acylated, hydroxylated, have disulfide bonds added, and undergo enzymatic cleavage.

List some common antibiotics that act on the bacterial 30S ribosomal subunit:
Aminoglycosides (ie, streptomycin, gentamicin, tobramycin, amikacin) are bactericidal and tetracyclines (ie, tetracycline, doxycycline, demeclocycline, minocycline) are bacteriostatic.

List some common antibiotics that act on the bacterial 50S subunit:
Chloramphenicol, erythromycin/ macrolides, lincomycin, clindamycin, streptogramins (ie, quinupristin, dalfopristin), and linezolid

What is the mechanism of action for chloramphenicol?
It inhibits the bacterial 50S peptidyl transferase—chloramphenicol is bacteriostatic.

What is the mechanism of action for erythromycin/macrolides?
They inhibit protein synthesis by blocking translocation via their binding to the 23S rRNA of the 50S ribosomal subunit—these antibiotics are bacteriostatic.

What is the mechanism of action for clindamycin?
It blocks peptide bond formation at the bacterial 50S ribosomal subunit— clindamycin is bacteriostatic.

CANCER DRUGS

What is the mechanism by which busulfan works?	Alkylating DNA, resulting in its deactivation
What is busulfan used to treat?	Chronic myelogenous leukemia (CML)
What are the major side effects associated with busulfan use?	Hyperpigmentation, pulmonary fibrosis
Describe the mechanism by which cyclophosphamide works:	Following hepatic bioactivation, it is an alkylating agent that cross-links DNA at guanine N-7.
What is cyclophosphamide used to treat?	Breast cancer, non-Hodgkin lymphoma
What are cyclophosphamide's toxicities?	Myelosuppression, hemorrhagic cystitis (preventable with the use of mesna, which reacts with the urotoxic metabolites)
What is the mechanism by which the nitrosoureas (ie, carmustine, lomustine, streptozocin) work?	Alkylating DNA
What are nitrosoureas used to treat?	Brain tumors (because they can cross the blood-brain barrier)
What is a common toxicity associated with the use of nitrosoureas?	Central nervous system (CNS), they can cause ataxia and/or dizziness
What is the mechanism by which cisplatin works?	Alkylating DNA
What is cisplatin used to treat?	Bladder, ovary, testicular, and lung small cell cancer
What are the common toxicities of cisplatin?	Renal toxicity and acoustic nerve damage
What is the mechanism of doxorubicin's action?	Intercalating DNA, thus creating a break and causing a decrease in replication while generating toxic radicals

What is doxorubicin used to treat?   Hodgkin lymphoma, sarcomas, and solid tumors (lung, ovary, breast, colon)

What are some toxicities of doxorubicin?   Cardiotoxicity, myelosuppression, alopecia

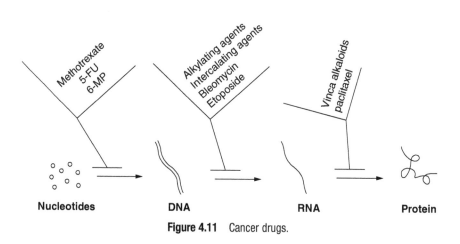

Figure 4.11   Cancer drugs.

SUMMARY CHART

Chart 4.1 Antibiotics

Antibiotic	Mechanism of Action
Chloramphenicol	Inhibits the bacterial 50S peptidyl transferase
Erythromycin/macrolides	Inhibits protein synthesis by blocking translocation via their binding to the 23S rRNA of the 50S ribosomal subunit
Clindamycin	Blocks peptide bond formation at the bacterial 50S ribosomal subunit

CLINICAL VIGNETTES—MAKE THE DIAGNOSIS

A 10-year-old white girl is brought in by her parents for evaluation of a skin disorder. The child has many freckles on her face, arms, and legs. In addition, she has telangiectasis along with areas of redness and hypopigmentation. The parents were told previously to limit her sun exposure as her skin is very sensitive to sunlight.

What disorder does this child appear to be suffering from?	Xeroderma pigmentosum
What is the inheritance pattern and etiology of this disorder?	Autosomal recessive; impaired nucleotide excision repair mechanism of ultraviolet light-damaged DNA bases
Why were this child's parents told to limit their daughter's exposure to sunlight?	Sunlight sensitivity can predispose to epidermoid and basal cell carcinoma

A 30-year-old white male presents with shortness of breath, coughing, and diminished exercise intolerance. He reports that the symptoms have gotten worse in the past 2 weeks. On physical examination, crackling sounds in the chest are heard. During the past medical history, the patient reveals that he was diagnosed with testicular cancer 5 years ago. His cancer went into remission after extensive chemotherapy. Chest x-ray and lung function testing indicate that the patient has developed pulmonary fibrosis.

What could be responsible for this patient's pulmonary fibrosis?	Bleomycin, a drug used to treat testicular cancer.
Name another condition that the aforementioned drug is used to treat:	Hodgkin lymphoma
What is the drug's mechanism of action?	Induction of DNA strand breaks via free radicals. It acts during the G_2 phase of the cell cycle.

A 55-year-old black male with high blood pressure and diabetes comes into your nephrology clinic. He has suffered from kidney problems for 10 years and has been on dialysis for the past 2 years. He wears a beeper and remains in contact with the transplant procurement team. The next night he is called and told that a compatible kidney had been found and he was chosen as its recipient.

Name one medication that should be administered following the kidney transplant:	Azathioprine, an immunosuppressant
Describe the drug's mechanism of action:	Azathioprine is a purine synthesis inhibitor, inhibiting the proliferation of cells, especially leukocytes.
What is the aforementioned drug derived from?	6-Mercaptopurine (6-MP)

A 19-year-old female presents with flu-like symptoms. She complains of fever, headaches, and sore throat. Directed questioning reveals that the patient had unprotected sex, 4 weeks back, with someone having HIV. Physical examination reveals cervical and axillary lymphadenopathy along with a nonspecific rash on her torso. The patient is concerned that she may have HIV.

What initial screening test should be done?	Enzyme-linked immunosorbent assay (ELISA)
Describe how this test would work in detecting HIV:	Patient's blood sample is probed with HIV antigen coupled to a color-generating enzyme to determine whether the patient's blood contains HIV antibodies. If the test solution has an intense color reaction, it indicates a positive test result.
If the first test result is positive, what test is commonly used to confirm the diagnosis?	Western blot
Describe the steps involved in the second test:	Electrophoresis for protein separation; identification by specific complexing with radio-labeled antibodies

A 4-year-old black girl is brought to the emergency department by her parents who say the child has been complaining of abdominal and flank pain for 1 month. Additionally, the parents notice blood in her urine. A CT scan of her abdomen demonstrates a mass lesion on her right kidney. The lesion is eventually biopsied and is revealed to be a Wilms tumor (nephroblastoma). A treatment strategy is developed, which includes nephrectomy and 18 weeks of chemotherapy. After starting her chemotherapy, the patient develops myelosuppression, detected by a complete blood count. Examination and questioning reveals more side effects, including constipation, and numbness and tingling in the hands and feet.

What class of chemotherapy agents was probably included in this patient's treatment regimen?	Vinca alkaloids (vincristine and vinblastine)
How do these drugs work?	Work in M phase of the cell cycle; bind to tubulin and block microtubule polymerization, thus not allowing the mitotic spindle to form
The neurotoxicity side effects seen in this patient are due to which specific drug?	Vincristine. Remember for the vinca alkaloid toxicities: vincristine—paralytic ileus and neurotoxicity, vinblastine—myelosuppression.

A 67-year-old Caucasian male presents to his dermatologist with a bump on his chin which occasionally bleeds. This lesion has been present for 5 years. The patient is a retired accountant and spends much of his free time fishing and golfing. The dermatologist biopsies this lesion and pathology reveals cancer.

Statistically, what type of cancer is this lesion most likely to be?	Basal cell carcinoma
Name a drug which is used to treat this cancer and acts upon the S phase of the cell cycle:	5-Fluorouracil
Name the major irreversible toxicity associated with its usage:	Myelosuppression
What autosomal recessive disease is associated with the development of skin cancer secondary to an inability to repair UV-induced DNA damage?	Xeroderma pigmentosum

A 23-year-old disheveled female presents to the emergency department complaining of chest pain and fatigue. Physical examination reveals fever, tachycardia, petechiae, splinter hemorrhages within the nailbeds and nontender maculae on her palms and soles (Janeway lesions). Further history reveals intravenous drug abuse. Blood cultures reveal *Staphylococcus aureus* bacteremia and she is diagnosed with endocarditis. She is started on ampicillin and gentamicin.

Gentamicin belongs to which class of antibiotics?	Aminoglycosides
How does gentamicin work?	Inhibits protein synthesis (acts on the 30S ribosomal subunit); bactericidal activity
What are the major side effects of gentamicin?	Nephrotoxicity (kidneys), hearing loss (ears), neuropathy (nerves), ie, **KEN**

A 45-year-old Native American female undergoes cholecystectomy for the treatment of symptomatic cholelithiasis. In the postoperative period, she becomes very ill and is transferred to the intensive care unit. She develops a severe wound infection and is found to have vancomycin-resistant enterococci in her blood and wound cultures. An infectious disease physician recommends the use of linezolid to treat this severe infection.

What is linezolid's mechanism of action?	Protein synthesis inhibitor (acts on the 50S ribosomal subunit)
What are the short-term side effects of linezolid?	Headache, diarrhea, nausea
What are the long-term side effects of linezolid?	Bone marrow suppression, thrombocytopenia (low platelet counts), peripheral neuropathy, lactic acidosis

A 37-year-old African American female presents to her primary care physician with a lump on her left breast, which she noticed while bathing. Her mother was diagnosed with breast cancer at 31 years of age. The patient has two daughters. Mammography demonstrates a mass at the upper outer quadrant of her left breast. The lump is excised and pathology reveals BRCA1-positive, HER2-positive invasive ductal carcinoma.

The BRCA1 gene encodes which type of protein?	DNA repair regulator
Why are germ line mutations in the BRCA1 or BRCA2 genes important to be tested for?	Associated with increased risk of developing breast cancer and/or ovarian cancer
What type of gene is HER2?	Proto-oncogene encoding for a human epidermal growth factor receptor, which may be overexpressed
What is the implication of HER2 positivity in breast cancer?	Associated with increased disease recurrence and worse prognosis
What drug can be used to treat HER2-positive breast cancer?	Trastuzumab, which binds to the overexpressed growth factor receptor

CHAPTER 5

Genetics

OVERVIEW

In autosomal dominant (AD) inheritance, are males or females more affected?	Both are equally affected
What genes are usually affected in disorders with AD inheritance?	Structural genes
Name some common AD disorders:	Huntington disease, familial hypercholesterolemia type IIa, neurofibromatosis I and II, tuberous sclerosis, adult polycystic kidney disease (APKD), Marfan syndrome, hereditary spherocytosis, familial adenomatous polyposis (FAP), von Hippel-Lindau syndrome, von Willebrand disease, all multiple endocrine neoplasia (MEN) syndromes, achondroplasia, familial hypocalciuric hypercalcemia

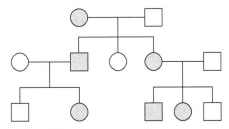

Figure 5.1   Autosomal dominant pedigree.

Two carriers of a lethal autosomal recessive (AR) disorder have a child. What is the probability that the child will die as a result of this disorder?	25%

What is the probability that their child will be healthy but also a carrier?   66%

What is meant by the term horizontal transmission?   A disease phenotype is seen in multiple siblings, usually with no earlier generations affected.

What is consanguinity?   The mating of related individuals, which may be suspected when analyzing the cause of rare AR disorders

Name some characteristics of AR traits/disorders: Often due to enzyme deficiencies, often more severe than AD disorders, and patients often present in childhood

Are males or females more commonly affected in AR disorders?   Males and females are equally affected

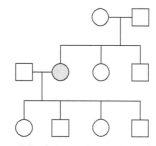

Figure 5.2   Autosomal recessive pedigree.

Are males or females more commonly affected in X-linked recessive disorders?   Males (remember, males only have one X-chromosome)

What is lyonization?   The normal phenomenon that wherever there are two or more haploid sets of X-linked genes in each cell, all but one of the genes are randomly inactivated and have no phenotypic expression

What is the importance of lyonization in X-linked disorders?   It explains the more variable expression of X-linked traits in women than in men.

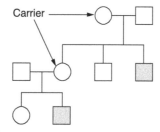

Figure 5.3   X-Linked recessive pedigree.

Name some common X-linked recessive disorders:	Fragile X, hemophilia A and B, Duchenne muscular dystrophy, Fabry, glucose-6-phosphate dehydrogenase deficiency, Hunter syndrome, Lesch-Nyhan syndrome, Bruton agammaglobulinemia, Wiskott-Aldrich syndrome
What is the main feature of mitochondrial inheritance?	Transmission of the disorder is only through the mother.

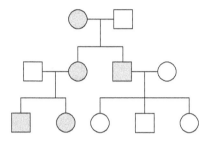

Figure 5.4 Mitochondrial inheritance pedigree.

What does variable expression mean?	The nature and severity of the phenotype varies from one individual to another.
What does incomplete penetrance mean?	Not all individuals with a mutant genotype will show the mutant phenotype.
What is pleiotropy?	The ability of a single allele to have more than one distinguishable effect
What is imprinting?	Differences in a phenotype depending on whether a mutation occurs in the mother's or father's alleles
Give two examples of imprinting:	1. Angelman syndrome (deletion in the maternal allele of chromosome 15q12) 2. Prader-Willi syndrome (deletion in the paternal allele of chromosome 15q12)
Name the clinical manifestations of Angelman syndrome:	Mental retardation, ataxia, seizures, strange affect, inappropriate laughter (hence, the happy puppet syndrome)
Name the clinical manifestations of Prader-Willi syndrome:	Hypotonia, poor infant feeding followed by childhood obesity, hyperphagia, delayed psychomotor development

What is anticipation?

The severity of a disorder worsens or the disorder manifests earlier in future generations.

What group of genetic disorders generally exhibit anticipation?

Trinucleotide repeat disorders such as Huntington disease

Where is the mutation in Huntington disease?

Chromosome 4p (remember "Hunting 4 food")

Name the trinucleotide repeat expansion that causes Huntington disease:

CAG, a codon for glutamine

Name some features of Huntington disease:

Autosomal dominant disorder showcasing anticipation, usually presenting in the ages of 20–50 years, and affected patients can develop depression, chorea, jerky hyperkinetic movements, caudate atrophy (disease progression is directly proportional to the size of the lateral ventricles)

What is loss of heterozygosity?

If an individual has a mutation or develops a mutation in a tumor suppressor gene, then the other allele must be deleted or mutated before a neoplasm can develop.

Do oncogenes show loss of heterozygosity?

No

Name some characteristics of X-linked dominant inheritance:

Can be transmitted through either parent, all female offspring of an affected father will also be affected, males are generally more severely affected than females

Name some X-linked dominant disorders:

Xg blood group, vitamin D-resistant (hypophosphatemic) rickets, Rett syndrome, pseudohypoparathyroidism, ornithine transcarbamylase deficiency

What is a dominant negative mutation?

The dominant allele, even in a heterozygote state, produces a nonfunctional protein that exerts its effect on the normal allele and does not allow the production of a functional protein (hence, its dominance over the functional allele).

What is linkage disequilibrium?	The tendency for certain alleles at two different loci on the chromosome to occur together more often than by chance.
What is mosaicism?	When different cells in the body express different alleles (ie, X-linked inactivation in which a certain subset of cells inactivates one X-chromosome and another subset of cells inactivates the other X-chromosome)
What is gene flow?	Exchange of genes among different populations
When is a concordance study of twins used?	To determine heritability of multifactorial traits
What is the equation used to determine this heritability?	Heritability = $C_{mz} - C_{dz}/1 - C_{dz}$ C_{mz} is the concordance in monozygotic twins. C_{dz} is the concordance in dizygotic twins.
What is the Hardy-Weinberg equation?	$p^2 + 2pq + q^2 = 1$ $p + q = 1$ p and q are separate alleles; 2pq = heterozygote prevalence
What are the four assumptions of Hardy-Weinberg population genetics?	1. No mutations 2. Random mating 3. No migration 4. No selection
What is the Hardy-Weinberg equation used to calculate in recessive disease models?	Carrier frequency
What can cause allele frequency to increase in a population?	Heterozygote advantage, genetic bottleneck, migration, new mutation

CHROMOSOMAL ABNORMALITIES

What is aneuploidy?	The state of having an abnormal number of chromosomes, not an exact multiple of the haploid number
What is polyploidy?	The state of having three or more haploid sets of chromosomes

What is the most common cause of polyploidy?	Polyspermy or egg fertilization by two sperms, accounts for two-thirds of the cases
What is a reciprocal translocation?	The mutual exchange of chromosomal material between two different chromosomes
Will an individual with a balanced reciprocal translocation be phenotypically normal?	Yes (but their offspring is at increased risk of having an unbalanced translocation)
What is a Robertsonian translocation?	A translocation in which the centromeres of two acrocentric chromosomes appear to have fused, forming an abnormal chromosome consisting of the long arms of two different chromosomes
Describe the characteristics of an acrocentric chromosome:	A chromosome with a peripherally placed centromere; therefore, one arm is shorter than the other
Which human chromosomes are acrocentric?	Chromosomes 13, 14, 15, 21, and 22
What is the difference between a terminal deletion and interstitial deletion?	A terminal deletion involves the end of a chromosome but an interstitial deletion involves a region within a chromosome.
What type of genetic mutation may result in trisomy of a particular chromosomal segment?	Duplication
What is the difference between a pericentric inversion and paracentric inversion?	A pericentric inversion includes the centromere and a paracentric inversion does not include the centromere.
What is a ring chromosome?	A chromosome with ends joined to form a circular structure; the ring form is abnormal in humans but the normal form of the chromosome in certain bacteria.
What percentage of live births have a chromosomal abnormality?	0.5%
What percentage of clinically recognized pregnancies result in spontaneous abortion?	15%

Chromosomal abnormalities are
implicated in what percentage of
spontaneous first-trimester abortions?

50%

Identify the following chromosomal
structural abnormalities given
in Fig. 5.5:

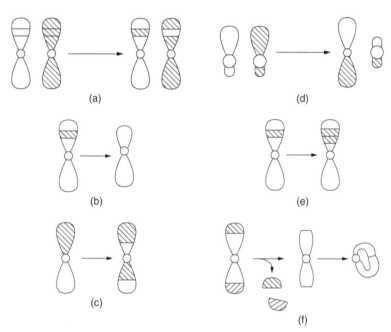

Figure 5.5

(a) Reciprocal translocation
(b) Deletion
(c) Duplication

(d) Robertsonian translocation
(e) Inversion
(f) Ring conversion

What disorder is characterized by
congenital mental retardation, flat facial
profile, duodenal atresia, and congenital
heart disease?

Down syndrome (trisomy 21)

This disorder is associated with what
maternal parameter?

Advanced maternal age

What is the most common cause
of this disorder?

Meiotic nondisjunction of homologous
chromosomes

What percentage of cases are a result
of unbalanced Robertsonian
translocations?

3%

What is the incidence of this disorder?	1:800
What disorder is characterized by severe mental retardation, clenched fists with overlapping digits, and rocker-bottom feet?	Edwards syndrome (trisomy 18)
What is the prognosis of an individual with this disorder?	Incompatible with life beyond 2 years of age
What is the incidence of this disorder?	1:8000
What disorder is characterized by severe mental retardation, microcephaly with microphthalmia, cleft palate/cleft lip, and polydactyly?	Patau syndrome (trisomy 13)
What is the prognosis of an individual with this disorder?	Death commonly by 1 year of age
What is the incidence of this disorder?	1:6000
Describe the symptoms of Klinefelter syndrome:	XXY male with testicular atrophy, gynecomastia, eunuchoid body shape, and female hair distribution
What is the cause of Klinefelter syndrome?	Meiotic nondisjunction of the sex chromosomes
What is the gonadal state of individuals with this disorder?	Hypogonadism (but at increased risk of developing breast cancer)
What is the incidence of this disorder?	1:850
What are the symptoms of Turner syndrome?	XO female with short stature, ovarian dysgenesis, webbing of the neck, and primary amenorrhea
Individuals with Turner syndrome are at increased risk to have what cardiac abnormality?	Coarctation of the aorta
What is the incidence of Turner syndrome?	1:3000
What are the symptoms of an individual with double Y?	XYY males who are phenotypically normal, but usually tall with severe acne
What is the incidence of this disorder?	1:1000 (increased frequency among inmates of penal institutions)

GENETIC DISORDERS

What disorder characterized by macro-orchidism, a long face with large jaw, large everted ears, and autism is the second most common cause of congenital mental retardation?

Fragile X syndrome

What genetic abnormality is associated with this disorder?

Progressive expansion of an unstable DNA triple repeat (CGG) on the X-chromosome

One would like to run an electrophoresis gel to separate the abnormal X-chromosome from the normal X-chromosome in a carrier female. Relative to the normal X-chromosome, would the abnormal X-chromosome run a longer or shorter distance on the gel?

Shorter distance because the expanded abnormal X-chromosome would be molecularly larger than the normal X-chromosome

What method is normally used in diagnosis?

Lymphocyte culturing in either folate-deficient medium or with chemical agents such as methotrexate; histological analysis revealing >4% of metaphase chromosomes with a characteristic break is diagnostic

Describe the inheritance pattern of this disorder:

X-linked recessive

What is the incidence of this disorder?

1:1500 males

What disorder is characterized by severe mental retardation, microcephaly, cardiac abnormalities, and laryngeal malformation?

Cri du chat syndrome (ie, cry of the cat)

Describe the genetic abnormality associated with this disorder:

Partial deletion of the short arm of chromosome 5 (46, XX or XY, 5p–)

What are the symptoms of von Recklinghausen disease (neurofibromatosis type I)?

Café-au-lait spots, neural tumors (optic gliomas and astrocytomas), Lisch nodules, and increased tumor susceptibility

This disorder is associated with a mutation on which chromosome?	Long arm of chromosome 17 (remember there are 17 letters in von Recklinghausen)
Describe the inheritance pattern of von Recklinghausen disease:	Autosomal dominant
What is the incidence of von Recklinghausen disease?	1:3000
What disease is characterized by cystic bilateral enlargement of the kidneys, hematuria, hypertension, and progressive renal failure?	Adult polycystic kidney disease (APKD)
This disorder is associated with a mutation on which chromosome?	*APKD1* gene on chromosome 16
What is the inheritance pattern of this disorder?	Autosomal dominant
What is the inheritance pattern of the juvenile form of this disorder?	Autosomal recessive
Marfan syndrome is caused by a mutation in the gene encoding for what protein?	Fibrillin (FBN), a component of connective tissue
The abnormality in Marfan syndrome is an example of what type of gene mutation?	Antimorphic gene mutation
Where is this gene located?	*FBN1* is located on chromosome 15
What are the clinical manifestations of Marfan syndrome?	Tall stature, long and thin fingers, scoliosis, subluxation of the lenses, and cystic medial necrosis of the aorta (leading to aortic incompetence and dissecting aortic aneurysms)
How is Marfan syndrome inherited?	Autosomal dominant pattern
What are the symptoms of velocardiofacial syndrome (VCFS)?	Cleft palate, mental retardation, hypernasal speech, and cardiac abnormalities
What other syndrome closely resembles VCFS?	DiGeorge syndrome, which has additional immune system deficiencies associated with lack of a thymus gland

What type of genetic mutation is associated with VCFS?

Microdeletion on chromosome 22q

What is the incidence of VCFS?

1:3000

CANCER GENETICS

Describe the multistage model for carcinogenesis:

Cancer is not a single event; rather, it is a multistage process in which an initial mutation in a somatic cell is followed by subsequent mutations and alterations which accumulate over time and lead to cancer.

What is a tumor suppressor gene?

A gene whose function is to suppress cellular proliferation

What is the role of a tumor suppressor gene in neoplastic transformation?

It suppresses neoplastic transformation; loss of a tumor suppressor gene (both alleles) through chromosomal aberration leads to heightened susceptibility to neoplastic changes.

Describe the Knudsen hypothesis:

An explanation for the bilateral (and earlier) occurrence of hereditary retinoblastoma; if one tumor suppressor gene is mutated by inheritance, only one somatic mutation is needed to inactivate the other allele. (Fig. 5.6)

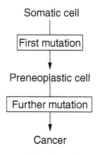

Figure 5.6 Knudsen hypothesis.

What tumor suppressor gene is implicated in the development of most human cancers?

p53

Where is this tumor suppressor gene located?	17p
What protein does it encode?	A transcription factor that regulates the expression of an inhibitor of cell cycle progression; when it is nonfunctional, cells with damaged DNA are allowed to progress through the cell cycle, possibly accumulating more damage, and predisposing these cells to neoplastic changes.
What disorder is a result of germ-line mutations in this tumor suppressor gene?	Li-Fraumeni syndrome
What cancers occur most commonly in individuals with this disorder?	Breast cancer, brain tumors, acute leukemia, soft tissue sarcomas, osteosarcoma, and adrenal cortical carcinoma
What disorder is characterized by hemangioblastomas of the retina, cerebellum, and medulla?	von Hippel-Lindau disease
This disorder is associated with the deletion of what gene?	*von Hippel-Lindau* tumor suppressor gene
Where is this gene located?	Chromosome 3p
What disorders are associated with a deletion in the *adenomatous polyposis coli (APC)* tumor suppressor gene?	Familial adenomatous polyposis (FAP), Gardner syndrome, some Turcot syndrome families, attenuated adenomatous polyposis coli (AAPC)
Where is the *APC* gene located?	Chromosome 5q
Describe the function of the *APC* gene:	Promotes apoptosis by sequestering the growth stimulatory effects of β-catenin
The *BRCA1* and *BRCA2* tumor suppressor genes are implicated in the development of what tumors?	Breast cancer, ovarian cancer
Where are these genes located?	*BRCA1* on chromosome 17q, *BRCA2* on chromosome 13q

These genes encode for what type of proteins?	DNA repair regulators
What is the lifetime risk of developing breast cancer for a female with a germ-line mutation of *BRCA1*?	80%
What is the lifetime risk of developing ovarian cancer for a female with a germ-line mutation of *BRCA1*?	60%
What is the lifetime risk of developing breast or ovarian cancer for a female with a germ-line mutation of *BRCA1*?	Almost 100%
Does a female with an individual germ-line mutation of *BRCA1* or *BRCA2* have a greater lifetime risk of developing ovarian cancer?	*BRCA2*
What is an oncogene?	A gene that codes for a protein involved in cell growth or regulation but may foster malignant processes if mutated or activated by contact with retroviruses
List the proteins that an oncogene may code for:	Protein kinases, guanosine 5′-triphosphatases (GTPases), nuclear proteins, growth factors
What is the Philadelphia chromosome?	An abnormal minute chromosome formed by a rearrangement of chromosomes 9 and 22; forms a *bcr-abl* fusion gene which translates for a protein with abnormal tyrosine kinase activity
The Philadelphia chromosome is found in cultured leukocytes of individuals with what disorder?	Chronic myelogenous leukemia (CML)
What drug is used to treat this disorder?	Imatinib
Describe Imatinib's mechanism of action:	Imatinib is a protein-tyrosine kinase inhibitor that inhibits the *bcr-abl* tyrosine kinase created by the Philadelphia chromosome; thus, it inhibits proliferation and induces apoptosis in *bcr-abl*-positive cell line.

What translocation is associated with Burkitt lymphoma?	t(8;14)
What gene is overactivated following this translocation?	*c-myc*, a transcription factor essential for mitosis of mammalian cells
What virus may be implicated in the development of Burkitt lymphoma?	Epstein-Barr virus
What translocation is associated with follicular lymphoma?	t(14;18)
What gene is overactivated following this translocation?	*bcl-2*, an inhibitor of apoptosis
What translocation is associated with acute myelogenous leukemia (AML) (M3 type)?	t(15;17)
AML is responsive to treatment with what drug?	All-trans retinoic acid
What translocation is associated with Ewing sarcoma?	t(11;22) (remember, Patrick Ewing wore number 11 + 22 = 33)
What gene is formed as a result of this translocation?	*EWS-FLI1* fusion gene, which transcribes an aberrant transcription factor
What translocation is associated with mantle cell lymphoma?	t(11;14)
What gene is overactivated following this translocation?	*cyclin D1*, a promoter of cell cycle progression
What is a carcinogen?	Any cancer-producing substance
What is the proposed mechanism of action for most carcinogens?	Cause damage to DNA and so generate mutations
What test identifies the mutagenicity of a potential carcinogen?	Ames test
Describe the test used in the question above:	A screening test for possible carcinogens using strains of *Salmonella typhimurium* that are unable to synthesize histidine; if the test substance produces mutations that regain the ability to synthesize histidine, the substance is carcinogenic (Fig. 5.7)

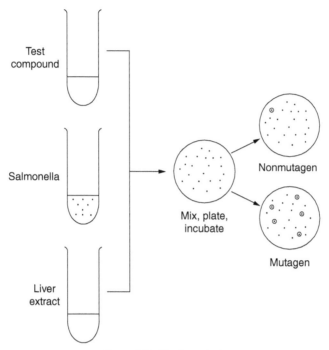

Figure 5.7 The Ames test.

List the classes of carcinogens:	Chemical carcinogens, viruses, radiation
List the associated carcinogens for the following organs:	
Liver	Aflatoxins, vinyl chloride, oral contraceptive pills, hepatitis B and C viruses, carbon tetrachloride
Esophagus	Nitrosamines, tobacco, gastroesophageal reflux disease (GERD)
Stomach	Nitrosamines
Lung	Tobacco, asbestos
Skin	Arsenic, radiation, human herpesvirus-8 (HHV-8) (Kaposi sarcoma)
Bladder	Naphthalene (aniline) dyes, *Schistosoma haematobium*
Cervix/penis/anus	Human papilloma virus
Blood	Human T-cell lymphotrophic virus type I

SUMMARY CHART

Chart 5.1 Cancer Genetics

Disease	Gene Involved	Translocation
Burkitt lymphoma	*c-myc*, transcription factor for mitosis	t(8;14)
Acute myelogenous leukemia (AML) (M3 type)	*RARα*, retinoic acid receptor	t(15;17)
Follicular lymphoma	*bcl-2*, inhibitor of apoptosis	t(14;18)
Ewing sarcoma	*EWS-FL1* fusion, acts as an aberrant transcription factor	t(11;22)
Mantle cell lymphoma	*Cyclin D1*, a promoter of cell cycle progression	t(11;14)

CLINICAL VIGNETTES—MAKE THE DIAGNOSIS

A 10-year-old White male is brought in by his parents for evaluation of his mental retardation. The boy has a long and narrow face, prominent ears, and enlarged testicles. The mother reports a possible history of mental retardation on her side of the family.

This child is likely suffering from what disorder?	Fragile X syndrome
What is the etiology behind the mutation responsible for this condition?	Triple-nucleotide repeat in the *FMR1* gene on the X-chromosome
What condition is more commonly the cause of mental retardation in males?	Down syndrome

A 39-year-old male complains of random episodes of unintentional movements. He reports that he developed an unsteady gait a few years ago, but it was not until recently that the uncoordinated, jerky body movements became apparent. Physical examination reveals slowed saccadic eye movements, physical instability, abnormal facial expression, and difficulty in speaking. Associated symptoms include sleep disturbances, weight loss, and mild memory deficits. He grew up as an adopted child and is unable to provide a complete family history.

What is the likely diagnosis for this patient?	Huntington disease
What is the inheritance pattern of this disorder?	Autosomal dominant
Neuroimaging would reveal what characteristic finding?	Atrophy of the striatum

A 40-year-old father presents with flank pain and hematuria. He denies a history of alcohol, tobacco, or drug use. Past medical history reveals nocturia, hypertension, and recurrent urinary tract infections. He reports that kidney problems run in his family; his father died from renal complications. On physical examination, his kidneys are palpable bilaterally and his blood pressure is 155/100 mm Hg. Serum creatinine levels are elevated. Renal ultrasound shows multiple masses in both kidneys.

What disorder is this patient suffering from?	Adult polycystic kidney disease
What chromosome is involved?	Chromosome 16
This disorder displays what mode of inheritance?	Autosomal dominant
Name one additional complication, not present in this patient, associated with this disorder:	Berry aneurysms

An 8-year-old girl is referred to your pediatric specialty service by her optometrist because of ectopia lentis (detached lens). The girl also complains of undue fatigue and back pain. She is tall for her age, with disproportionately long, slender limbs. Physical examination shows pectus excavatum, arachnodactyly, and hypermobility of the joints. Serum and urine analysis tests are normal.

What is the likely diagnosis for this patient?	Marfan syndrome
How is this disorder inherited?	Autosomal dominant pattern
This disease is caused by a mutation in what gene?	Gene encoding fibrillin, a connective tissue protein
What is the most frequent cause of death in these patients?	Acute rupture of an aortic aneurysm

A 19-year-old woman presents to the emergency room with progressive vision loss. She reports that her vision has become increasingly blurry in the past 2 weeks, and she feels that her eyes are bulging outward. Physical examination is notable for the following: more than six light brown spots on the skin, freckling of the axilla, pigmented nodules on the surface of the iris, and multiple subcutaneous lumps. MRI confirms the presence of optic gliomas bilaterally.

What is the underlying disorder of this patient?	Neurofibromatosis type I (von Recklinghausen)
What is the inheritance pattern of this disease?	Autosomal dominant
The mutated gene associated with this disorder is located on what chromosome?	Chromosome 17 (remember, count the letters in von Recklinghausen)

A 15-year-old girl comes into the pediatric clinic with primary amenorrhea. She is concerned with her delay in pubertal onset and feels that she doesn't look like other girls. The patient's height is 4'8, and her neck appears to be webbed. Physical examination reveals a lack of breast development, a broad chest, cubitus valgus, and low-set ears.

What is the clinical diagnosis for this patient?	Turner syndrome
What molecular test would be done to confirm the diagnosis?	Chromosome analysis (karyotyping).
The aforementioned test would likely show what results?	45 chromosomes, with only one X chromosome

A 28-year-old male complains of vision loss and difficulty walking. The patient also reports that he has been suffering from headaches and coordination problems. Family history reveals that the patient's father died of cancer at an early age. Past medical history reveals that the patient has undergone surgical resection for a pheochromocytoma. MRI scans indicate the presence of hemangioblastomas on the cerebellum and the retina.

What is the underlying disorder for this patient?	von Hippel-Lindau disease
Describe the etiology behind this disease:	Deletion of the *von Hippel-Lindau* tumor suppressor gene on chromosome 3p
This disorder is inherited in what pattern?	Autosomal dominant
What is the leading cause of death in these patients?	Renal cell carcinoma

A 48-year-old male complains of left upper quadrant abdominal pain, tiredness, and weight loss due to early satiety. Physical examination confirms splenomegaly and reveals modest enlargement of the liver and lymph nodes. Blood tests reveal a markedly elevated white blood cell count. Leukocyte cultures are observed under a microscope, and an abnormal chromosome is detected. The abnormal chromosome looks like a short chromosome 22.

Name the likely diagnosis for this patient:	Chronic myelogenous leukemia. The abnormal chromosome (called the Philadelphia chromosome) is a hallmark finding for CML.
What is the initial treatment for this disorder?	Imatinib
How does this drug work?	It inhibits the protein formed by the *bcr-abl* fusion gene in the Philadelphia chromosome. The Philadelphia chromosome is the product of a reciprocal translocation of chromosomes 9 and 22.

CHAPTER 6

Cell Biology and Physiology

PLASMA MEMBRANE

List the biological ions important to cell physiology:	Sodium, potassium, magnesium, calcium, hydrogen, chloride
What feature of a cell allows for the compartmentalization of biological functions?	Lipid membranes
What substances are the typical constituents of a plasma membrane?	Cholesterol, phospholipids, sphingolipids, glycolipids, proteins
The plasma membrane has what basic molecular structure?	Phospholipid bilayer
What are two important characteristics of a phospholipid?	1. Polar head 2. Hydrophobic tail
In what organelle are phospholipids synthesized?	Endoplasmic reticulum

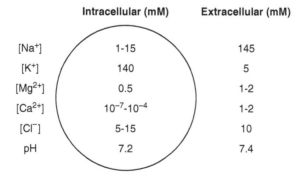

	Intracellular (mM)	Extracellular (mM)
$[Na^+]$	1-15	145
$[K^+]$	140	5
$[Mg^{2+}]$	0.5	1-2
$[Ca^{2+}]$	$10^{-7}-10^{-4}$	1-2
$[Cl^-]$	5-15	10
pH	7.2	7.4

Figure 6.1 Concentrations of ions.

What structure(s) do phospholipids spontaneously form when placed in water?	Micelles or bilayers
The aforementioned structure is formed as a result of what property of water?	Polarity of water; hydrocarbons, such as phospholipids, placed in water disrupt the ability of water molecules to form hydrogen bonds but bilayer formation allows water to reform these hydrogen bonds
Sphingolipids are an important constituent of what type of tissue?	Nerve tissue (see the lysosomal storage diseases section)
What chemical force drives the formation of a lipid membrane?	Hydrophobic interactions between the phospholipid tails; hydrophilic interactions between phospholipids heads and extracellular/cytosolic fluid
What are the four major phospholipids present in the plasma membrane?	1. Phosphatidylserine 2. Phosphatidylinositol 3. Phosphatidylcholine 4. Phosphatidylethanolamine
What two types of lipids are more predominant on the outer leaflet of the bilayer?	1. Phosphatidylcholine (lecithin) 2. Sphingomyelin
What two types of lipids are more predominant on the inner leaflet of the bilayer?	1. Phosphatidylserine 2. Phosphatidylethanolamine
The sugar component of a glycoprotein or glycolipid is usually found on which side of the plasma membrane?	Extracellular side
List the molecules and structures in which lecithin is a component:	Surfactant (ie, dipalmitoyl phosphatidylcholine), myelin, red blood cell (RBC) membranes, bile
Deficiency of surfactant is associated with what potentially fatal neonatal disorder?	Neonatal respiratory distress syndrome (hyaline membrane disease)
What is the treatment of choice for the aforementioned disorder?	Maternal prenatal administration of betamethasone or artificial surfactant for the infant
What are the two factors that influence the fluidity of the plasma membrane?	1. Cholesterol content 2. The length of the unsaturated fatty chains of the phospholipids
Fluidity of the plasma membrane is crucial to what cellular processes?	Exocytosis, endocytosis, membrane trafficking, membrane biogenesis

What groups of molecules can freely diffuse across a lipid membrane?

Hydrophobic molecules (ie, O_2, CO_2, N_2, benzene) and small uncharged polar molecules (ie, H_2O, urea, glycerol)

Describe what happens to a cell placed in the following solutions:

Hypertonic solution?

Decreases in size

Isotonic solution?

No change in size

Hypotonic solution?

Increases in size

The influx or efflux of molecules across a lipid membrane is dependent on what two factors?

1. Permeability of the lipid membrane
2. Driving force

The driving force is directly proportional to what factors?

For uncharged molecules, driving force is directly proportional to the concentration gradient across the lipid membrane. For charged molecules, the driving force is directly proportional to the concentration gradient and electrochemical gradient across the lipid membrane.

MEMBRANE PROTEINS

Name the three classes of membrane proteins:

1. Integral proteins (including transmembrane proteins)
2. Surface membrane proteins
3. Inner membrane proteins

Identify the membrane proteins in Fig. 6.2.

(a) Integral proteins
(b) Peripheral proteins
(c) Oligosaccharide
(d) Carbohydrate

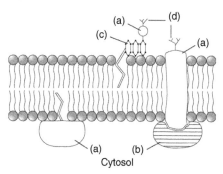

Figure 6.2

Which membrane proteins stabilize the plasma membrane and maintain RBC shape?

Integrins (ie, spectrin, actin, band 4.1 protein, ankyrin)

What disease results from defects in spectrin?

Hereditary spherocytosis

Transport proteins belong to which class?

Integral proteins

What role may surface membrane or inner membrane proteins play in transport?

They can associate with transport proteins to modulate their activity.

Can membrane proteins alter their intracellular/extracellular face orientation?

No

Can membrane proteins alter their lateral orientation within the plasma membrane?

Yes

What factors dictate the maximum rate of facilitated diffusion?

Number of transporters, rate of channel action, affinity of the transporter for the molecule, concentration of the molecule

What transport proteins are responsible for the facilitated diffusion of charged molecules across a lipid membrane?

Carrier proteins (more selective, slower rate of transport) and channel proteins (less selective, faster rate of transport)

Define passive transport:

Movement of a molecule/ion down its respective concentration/electrochemical gradient (requires no expenditure of energy)

Define active transport:

Movement of a molecule/ion against its respective concentration/electrochemical gradient (requires expenditure of energy)

Which type of transport proteins can carry out active transport?

Only carrier proteins

Where do primary transporters allocate the energy required for active transport?

Adenosine triphosphate (ATP) hydrolysis

How do secondary transporters allocate the energy required for active transport?

They couple the movement of a molecule against its concentration gradient to the movement of another molecule down its concentration gradient.

Identify the channel proteins on Fig. 6.3:

(a) Simple diffusion
(b) Ion channel-mediated diffusion
(c) Carrier protein-mediated diffusion

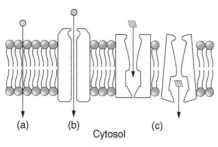

(a)   (b)   (c)
          Cytosol

Figure 6.3

Channel proteins typically transport what type of molecules?	Small, inorganic ions
The gating mechanism determines what functional property of a channel protein?	Permeability
Gating mechanisms are classified into what three categories?	1. Voltage gated 2. Ligand gated 3. Mechanically gated mechanisms
Carrier proteins typically transport what type of molecules?	Although more selective, they may transport any type of molecule
What are the four major classes of carrier proteins?	1. Uniporters 2. Antiporters 3. Symporters 4. Ion-transporting ATPases

NA⁺/K⁺ ATPASE

List some key cell functions in which maintenance of low Na⁺ and high K⁺ concentrations intracellularly are important:	Regulation of voltage, regulation of pH, membrane potential, transport of metabolites, water balance
Where in the cell is the Na⁺/K⁺ ATPase located?	Exclusively in the plasma membrane of all cell types (especially in neurons)

What is the net influx and efflux of Na⁺ and K⁺ per catalytic cycle?

For each ATP consumed, 3Na⁺ leave and 2K⁺ enter the cell

How many subunits does the Na⁺/K⁺ ATPase have?

Tetrameric; two α-subunits which contain the binding sites for Na⁺/K⁺, ATP and ouabain, and two glycosylated β-subunits which play a role in maturation and localization

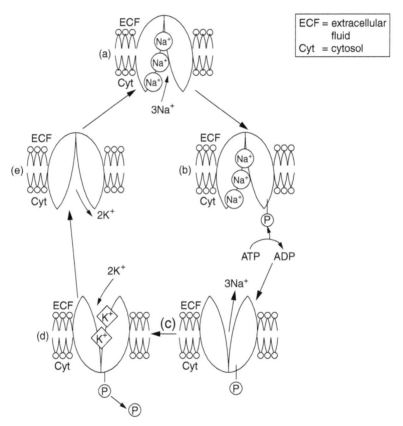

Figure 6.4 Na⁺/K⁺ ATPase. (a) 3Na⁺ from cytosol into Na⁺/K⁺ ATPase; (b) 3Na⁺ inside Na⁺/K⁺ ATPase, ATP degraded to ADP, phosphate group binds Na⁺/K⁺ ATPase; (c) 3Na⁺ from Na⁺/K⁺ ATPase into extracellular space; (d) 2K⁺ from extracellular space into Na⁺/K⁺ ATPase, phosphate group unbinds from Na⁺/K⁺ ATPase; (e) 2K⁺ from Na⁺/K⁺ ATPase into cytosol.

What is meant by the term electrogenic transporter?

Activity of the transporter creates an electrical current

What percentage of all metabolic energy in a resting organism is directed at fueling the Na⁺/K⁺ ATPase?

30%

Describe the mechanism by which digitalis (ouabain) increases cardiac contractility:

Digitalis competes with K^+ for binding on the extracellular surface of the transporter. This shuts off the phosphatase activity and interrupts the phosphorylation/dephosphorylation cycle, which leads to an increased Na^+ concentration intracellularly. Without a Na^+ concentration/electrochemical gradient, the Na^+/Ca^{2+} antiport loses functionality and results in an increased intracellular concentration of Ca^{2+}. Greater intracellular Ca^{2+} concentration causes stronger and longer contractions of cardiac muscle.

For what electrolyte must the dosage of digitalis (ouabain) be controlled?

K^+

MAJOR ION TRANSPORT PROTEINS

List some key cell functions in which Ca^{2+} plays a major role:

Contraction of all types of muscle, secretion of hormones and neurotransmitters, cell division, directed migration of nonmuscle cells, processing of visual pigments, apoptosis, and acute regulation of carbohydrate metabolism

What Ca^{2+}-binding protein is found in all cell types?

Calmodulin

What are the two types of Ca^{2+} transport channels?

1. Voltage-gated Ca^{2+} channels
2. Ligand-gated Ca^{2+} channels

Which type of Ca^{2+} channel is important in the contraction/relaxation cycle of cardiac muscle and the constriction/dilation cycle of vascular smooth muscle?

Voltage-gated Ca^{2+} channels (cardiac and smooth muscle cells have slow Ca^{2+} channels but skeletal muscle does not)

What drugs target the aforementioned Ca^{2+} channels?

Dihydropyridine Ca^{2+} blockers (ie, verapamil, diltiazem)

Which type of Ca^{2+} channel is found in the sarcoplasmic reticulum of striated muscle cells?

Ligand-gated Ca^{2+} channels

Where in the cell is the Na^+/Ca^{2+} antiporter located?

In the plasma membrane

What is the main role of the Na⁺/Ca²⁺ antiporter?

To remove excess cytosolic Ca^{2+}

The Na⁺/Ca²⁺ antiporter is ultimately dependent on what other transporter in order to function?

Na^+/K^+ ATPase; the Na^+ gradient created by the Na^+/K^+ ATPase produces an influx of Na^+ which is the driving force behind the eventual efflux of Ca^{2+}

List the three types of Ca²⁺ ATPases along with their regulatory molecules:

1. Plasma membrane form (calmodulin regulated)
2. Endoplasmic/sarcoplasmic reticulum form (not regulated)
3. Cardiac sarcoplasmic reticulum form (phospholamban regulated) (Fig. 6.5)

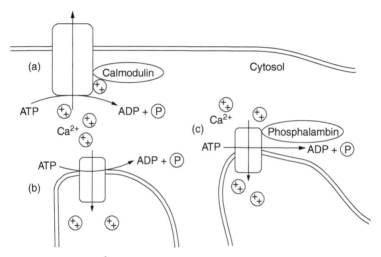

Figure 6.5 Three types of Ca^{2+} ATPases. (a) Plasma membrane form; (b) Endoplasmic/sarcoplasmic reticulum form; (c) Cardiac sarcoplasmic reticulum form.

What is the ideal intracellular pH?

7.1

Why does H⁺ move into the cell (remember the ideal extracellular pH is 7.4)?

Negative membrane potential creates an electrochemical gradient causing H^+ influx

What are the typical pH values in subcellular organelles such as the mitochondria and lysosomes?

Mitochondrial pH is 8, lysosomal pH is 5

Describe the activity of the Na⁺/H⁺ antiporter:

The influx of Na^+ is used to efflux H^+ in a 1:1 ratio.

What activates the Na$^+$/H$^+$ antiporter?	Drop in pH (as in glycolysis), growth factors, protein kinase C (PKC), oncogenes
What inhibits the Na$^+$/H$^+$ antiporter?	Amiloride, cyclic adenosine monophosphate (cAMP)-dependent protein kinase
Why is the Na$^+$-HCO3$^-$/H$^+$-Cl$^-$ exchanger twice as efficient as the Na$^+$/H$^+$ antiporter in alkalinizing the intracellular compartment?	Per cycle this exchanger removes H$^+$ and brings in HCO$_3^-$, which can neutralize the acidic pH
What drug can inhibit the Na$^+$-HCO$_3^-$/H$^+$-Cl$^-$ exchanger?	Stilbenedisulfonates
What types of cells necessary for the transport of CO$_2$ contain carbonic anhydrase?	RBCs
What is the major role of the Cl$^-$/HCO$_3^-$ exchanger?	To eliminate alkalinity in the intracellular compartment (activated when the intracellular pH rises above 7.2–7.4)

Systemic capillaries (↑pCO$_2$, ↓pO$_2$)   Pulmonary capillaries (↓pCO$_2$, ↑O$_2$)

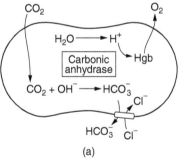 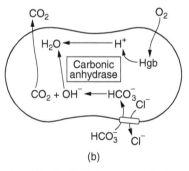

(a)   (b)

Figure 6.6   CO$_2$ Transport in RBCs. (a) Systemic capillaries; (b) Pulmonary capillaries.

What cells of the gastrointestinal (GI) tract express the H$^+$ ATPase?	Parietal cells of the stomach
What syndrome of recurrent stomach and duodenal ulcers is caused by a gastrin-secreting tumor?	Zollinger-Ellison syndrome
What class of drug is used to treat this disorder?	H$^+$ ATPase/proton pump inhibitors (ie, omeprazole, lansoprazole)

The Na^+, K^+, $2Cl^-$ cotransporter has the net effect of increasing the concentration of what ion intracellularly?	Cl^-
Where in the nephron can the Na^+, K^+, $2Cl^-$ cotransporter be found?	Luminal surface of the thick ascending loop of Henle (where NaCl is reabsorbed)
What disorder, characterized by secondary hyperaldosteronism, hypokalemic alkalosis, and growth retardation, may be associated with an autosomal recessive defect in the Na^+, K^+, $2Cl^-$ cotransporter?	Bartter syndrome
What drug used to treat hypertension inhibits the Na^+, K^+, $2Cl^-$ cotransporter?	Furosemide
What factor is important in the regulation of Cl^- channels?	Concentration of other ions (ie, Ca^{2+})
What protein is needed to phosphorylate, and thus activate the Cl^- channel?	Protein kinase A (PKA)
What disease characterized by recurrent pulmonary infections, pancreatic insufficiency, and bronchiectasis is a result of a defective chloride transporter?	Cystic fibrosis
What gene is mutated in this disorder?	*Cystic fibrosis transmembrane regulator (CFTR)*, a chloride ion channel found in epithelial cell membranes

NUCLEUS

Do eukaryotic or prokaryotic cells typically have a nucleus?	Eukaryotic cells
What cells found in the human body do not have a nucleus?	RBCs
What forms of RNA are synthesized in the nucleus?	All forms (ie, ribosomal RNA, messenger RNA, transfer RNA [rRNA, mRNA, tRNA])

Identify the labels on the nucleus in Fig. 6.7.

(a) Nuclear pore
(b) Euchromatin
(c) Heterochromatin
(d) Nucleolus
(e) Nuclear membrane.

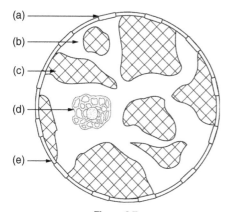

(a)
(b)
(c)
(d)
(e)

Figure 6.7

What structures constitute the nuclear envelope?	Two parallel membranes with an intervening perinuclear cisterna
What is the fibrous protein meshwork that coats the inner nuclear membrane and plays a role in structurally organizing the nucleus?	The nuclear lamina
What is the role of a nuclear pore?	To allow for the exchange of molecules between the nucleus and cytoplasm
What nuclear inclusion is involved in the synthesis of rRNA and its assembly into ribosome precursors?	The nucleolus
In what phase of the cell cycle does the nucleolus disappear?	Mitosis
What is the main role of chromatin?	RNA synthesis
Name the two forms of chromatin:	1. Heterochromatin 2. Euchromatin
Which form of chromatin corresponds to the Barr body in female mammalian cells?	Heterochromatin

Name a disorder in which no Barr body would be seen in the nucleus of a female mammalian cell:	Turner syndrome (XO)
Name a disorder in which a Barr body would be seen in the nucleus of a male mammalian cell:	Klinefelter syndrome (XXY)
During mitosis, what term describes division of the nucleus and of the cytoplasm?	Karyokinesis; cytokinesis
What form of cell division produces a haploid chromosome number?	Meiosis

ENDOPLASMIC RETICULUM

The endoplasmic reticulum is continuous with what nuclear structure?	Outer nuclear membrane
The rough endoplasmic reticulum synthesizes which proteins?	Membrane-packaged proteins including secretory, plasma membrane, and lysosomal proteins
What molecule may be added to proteins in the rough endoplasmic reticulum?	N-linked oligosaccharides
List the cell types with abundant rough endoplasmic reticulum:	Cells which produce protein in excess such as mucin-secreting goblet cells, antibody-secreting plasma cells
Describe the role of the smooth endoplasmic reticulum:	Steroid synthesis, detoxification of drugs, muscle contraction, and relaxation
What cell types are rich in smooth endoplasmic reticulum?	Leydig cells, adrenal cortex cells, hepatocytes, skeletal muscle cells

MITOCHONDRIA

The outer and inner membranes subdivide mitochondria into what two compartments?	1. Intermembrane compartment 2. Matrix compartment
Describe some of the important contents of mitochondria:	Enzymes of the tricarboxylic acid (TCA) cycle, enzymes of the electron transport chain, circular DNA, mRNA, tRNA, rRNA

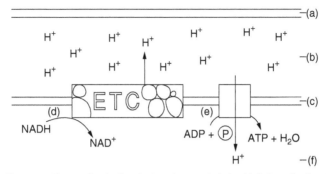

Figure 6.8 Chemosmotic coupling in the electron transport chain. (a) Outer mitochondrial membrane; (b) Intermitochondrial membrane space; (c) Inner mitochondrial membrane; (d) NADH dehydrogenase; (e) Mitochondrial ATPase; (f) Mitochondrial matrix.

What is the most important role of mitochondria?	ATP synthesis
What cells in the body do not contain mitochondria?	RBCs
Name two mechanisms by which mitochondria produce ATP:	1. Via the electron transport chain 2. Via the TCA cycle (oxidation of fatty acids, amino acids, and glucose)
During electron transport, H$^+$ ions are pumped from where to where?	H$^+$ ions are pumped from the matrix compartment across the inner mitochondrial membrane into the intermembrane compartment
What substances can directly inhibit electron transport?	Antimycin A, CN$^-$, CO, rotenone
What happens to the H$^+$ gradient when electron transport is inhibited?	Decreases
What are possible treatments for poisoning with electron transport inhibitors?	Supplemental O$_2$, amyl/sodium nitrate (CN$^-$ scavenger), methemoglobin
By what mechanism does oligomycin poison oxidative phosphorylation?	Directly inhibiting mitochondrial ATPase
What is the effect of oligomycin on the H$^+$ gradient?	Increases, because H$^+$ is pumped into the intermembrane compartment but it is sequestered there
How does 2,4-dinitrophenylhydrazine (DNP) poison oxidative phosphorylation?	Uncouples oxidative phosphorylation by increasing the permeability of the inner mitochondrial membrane

What is the effect of uncoupling on the H⁺ gradient and on O₂ consumption?

Decreases the H^+ gradient because H^+ pumped into the intermembrane compartment travels back into the matrix compartment without being coupled to ATP production; in order to compensate for the decreased H^+ gradient, electron transport continues and O_2 consumption is increased

What type of cell uses a mechanism similar to that of 2,4-DNP poisoning to produce heat?

Brown fat cells have a special transport protein in the inner mitochondrial membrane that uncouples respiration from ATP synthesis to produce heat.

Mitochondrial myopathies are transmitted through which parent?

Mother (all offspring of affected females may show signs of disease)

GOLGI APPARATUS

List the functions of the Golgi apparatus:

Distributing proteins and lipids

Modifying N-oligosaccharides on asparagine residues

Adding O-oligosaccharides to serine and threonine residues

Assembling proteoglycans from proteoglycan core proteins

Adding sulfur to sugars in proteoglycans and tyrosine residues in proteins

Adding mannose-6-phosphate to specific lysosomal proteins

Recycling and redistributing membranes

With what organelle is the *cis*-Golgi network associated?

Rough endoplasmic reticulum

How are proteins transferred to the Golgi apparatus?

Vesicles bud off from the rough endoplasmic reticulum and fuse with the *cis*-Golgi network

What is the role of the *cis*-Golgi network in protein processing?

Involved in protein sorting and retrieval

What is the role of the *trans*-Golgi network in protein processing?

Sorting proteins for their final destinations

Is protein transfer in the Golgi apparatus an energy-requiring process?	Yes, therefore it is dependent on ATP production
What organelle is responsible for the degradation of endocytosed material?	Lysosome
Describe the pathophysiology behind the inclusion vesicles associated with I-cell disease:	Failure of mannose-6-phosphate addition to lysosomal enzymes causes these proteins to be secreted instead of being delivered to the lysosomes. Thus, lysosomes cannot degrade the contents of these vesicles.
Name two types of coated vesicles:	1. Clathrin-coated vesicles 2. Coatamercoated vesicles
In which type of vesicle is a cage-like lattice formed around the vesicle?	Clathrin-coated vesicle
Which type of coat protein is formed by COPs?	Coatamer
Clathrin-coated vesicles are involved in what cellular functions?	Receptor-mediated uptake (endocytosis) of specific molecules by the cell, signal-directed (regulated) transport of proteins from the *trans*-Golgi network to lysosomes or secretory granules
Coatamer-coated vesicles are involved in what cellular processes?	Constitutive protein transport within the cell, anterograde transport of molecules from the rough endoplasmic reticulum to the Golgi apparatus (COP I and COP II proteins), retrograde transport of molecules from the Golgi apparatus to the rough endoplasmic reticulum (COP I proteins only)
How do lysosomes maintain their acidic pH?	ATP-powered H^+ pumps in their membranes
What role does the acidic lysosomal pH play in the recycling of vesicles?	Aids in the uncoupling of receptors and ligands, which allows the receptors to return to the plasma membrane and ligands to move to a late endosome
Do early or late endosomes have a more acidic pH?	Late endosomes (pH = 5.5 versus early endosomes pH = 6)
What proteins are found in late endosomes?	Lysosomal hydrolases, lysosomal membrane proteins

What constitutes a phagolysosome?   Fusion of a phagocytic vacuole with a late endosome or lysosome

What constitutes an autophagolysosome?   Fusion of an autophagic vacuole (carrying cell components targeted for destruction) with a late endosome or lysosome

What constitutes a multivesicular body?   Fusion of an early endosome containing endocytic vesicles with a late endosome

List the oxidative enzymes that can be found in peroxisomes: D-amino acid oxidase, urate oxidase, catalase

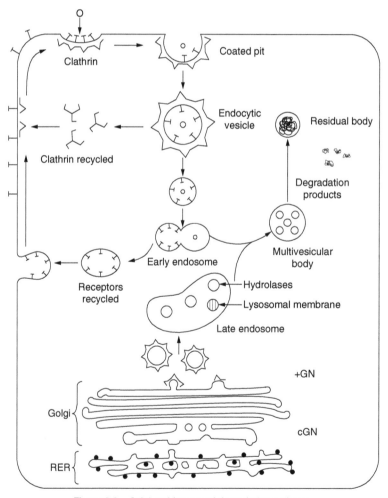

Figure 6.9   Golgi and lysosomal degradation pathway.

Defective activity of what enzyme(s) can cause chronic granulomatous disease?

Nicotinamide adenosine dinucleotide phosphate (NADPH) oxidase, lysosomal lysozymes, hydrolytic enzymes

What test is used to confirm the diagnosis of chronic granulomatous disease?

Nitroblue tetrazolium dye reduction test (negative result confirms disease)

LYSOSOMAL STORAGE DISEASES

What are the symptoms of Fabry disease?

Peripheral neuropathy of the hands and feet, angiokeratomas, cardiovascular and renal disease

What is the deficient enzyme in Fabry disease?

α-Galactosidase A

What substrate is ultimately increased due to the enzyme deficiency in Fabry disease?

Ceramide trihexoside (glycosphingolipid)

What is the inheritance pattern of Fabry disease?

X-linked recessive

What are the symptoms of Krabbe disease (globoid leukodystrophy)?

Growth retardation/difficulty feeding (remember "crabby baby"), peripheral neuropathy, hyperactive reflexes, developmental delay, and optic atrophy

What is the deficient enzyme in Krabbe disease?

Galactosylceramidase

What substrate is ultimately increased in Krabbe disease?

Galactocerebroside

What is the characteristic histological finding in Krabbe disease?

Multinucleated globoid macrophage cells

What is the inheritance pattern of Krabbe disease?

Autosomal recessive

What are the symptoms of Gaucher disease type I?

Massive splenomegaly, *Erlenmeyer flask* bone lesions and pathologic fractures, pancytopenia, thrombocytopenia; life span is not affected

What are the symptoms of Gaucher disease type II?

Type II is the acute neuronopathic form; hepatosplenomegaly, central nervous system (CNS) involvement, convulsions, and mental deterioration; life span is severely affected, with many dying at a young age

What is the deficient enzyme in Gaucher disease types I and II?

Glucocerebrosidase

What substrate is ultimately increased in Gaucher disease types I and II?

Glucocerebroside

What characteristic finding is seen histologically with Gaucher disease types I and II?

p-Aminosalicylic acid (PAS)-positive cellular inclusions

What inheritance pattern is seen in Gaucher disease types I and II?

Autosomal recessive

What are the symptoms of Niemann-Pick disease?

Delayed development, decreased visual acuity, diffuse neuronal involvement (cell death and cerebral atrophy), massive hepatosplenomegaly, infiltration of bone marrow, death by 3 years of age

What is the deficient enzyme in Niemann-Pick disease?

Sphingomyelinase

What substrate is ultimately increased in Niemann-Pick disease?

Sphingomyelin, with build-up of sphingomyelin and cholesterol in reticuloendothelial and parenchymal cells and tissues

What inheritance pattern is seen in Niemann-Pick disease?

Autosomal recessive

What sign in Niemann-Pick disease is also seen in Tay-Sachs disease?

Cherry-red spot on the macula

In what ethnicity is Niemann-Pick disease seen more often?

Ashkenazi Jews (similar to Tay-Sachs disease)

What are the symptoms of Tay-Sachs disease?

Motor and mental retardation starting at about 6 months of age, blindness, hyperacute hearing, death usually by 2-3 years of age

What is the deficient enzyme in Tay-Sachs disease?

Hexosaminidase A

What substrate is ultimately increased in Tay-Sachs disease?	G_{M2} ganglioside
What characteristic findings are seen histologically with Tay-Sachs disease?	Zebra bodies
What inheritance pattern is seen in Tay-Sachs disease?	Autosomal recessive
What are the symptoms of metachromatic leukodystrophy?	Spasticity with demyelinization and gliosis, increased cerebrospinal fluid (CSF) protein, ataxia, and dementia
What is the deficient enzyme in metachromatic leukodystrophy?	Arylsulfatase A
What substrate is ultimately increased in metachromatic leukodystrophy?	Sulfatide, with accumulation in the brain, kidney, liver, and peripheral nerves
What inheritance pattern is seen in metachromatic leukodystrophy?	Autosomal recessive
What are the symptoms of Hurler syndrome?	Coarse facial features (ie, gargoylism), hepatosplenomegaly, lesions of cardiac valves, narrowing of the coronary arteries, joint stiffness, kyphoscoliosis, mental retardation, and corneal clouding
What is the deficient enzyme in Hurler syndrome?	α-l-Iduronidase
What substrates are ultimately increased in Hurler syndrome?	Heparan sulfate and dermatan sulfate
What inheritance pattern is seen in Hurler syndrome?	Autosomal recessive
What are the symptoms of Hunter syndrome?	Mild form of Hurler syndrome, aggressive behavior, and no corneal clouding
What is the deficient enzyme in Hunter syndrome?	Iduronate sulfatase
What substrates are ultimately increased in Hunter syndrome?	Heparan sulfate and dermatan sulfate
What inheritance pattern is seen in Hunter syndrome?	X-linked recessive (therefore seen more in males than Hurler)

Of the lysosomal storage disorders mentioned, which are inherited in an X-linked recessive pattern?	Fabry and Hunter syndromes
How are the other disorders inherited?	Autosomal recessive

CYTOSKELETON

List the functions of the cytoskeleton:	Maintain cell shape, stabilize cell attachments, facilitate endo/exocytosis, promote cell motility
What are the three types of cytoskeletal filaments?	1. Microtubules 2. Microfilaments 3. Intermediate filaments
Describe the structure of a microtubule:	Helical array of polymerized α- and β-tubulin (13 per circumference)
The microtubule-organizing center is also known as what structure?	Centrosome
During cell division, where do microtubules attach on a chromosome?	Centromere
What proteins aid in the polymerization of tubulin?	MAPs (microtubule-associated proteins), tau proteins
What disease results from impaired polymerization of tubulin in leukocytes?	Chediak-Higashi syndrome
Describe the pathophysiology behind the neurofibrillary tangles associated with Alzheimer disease:	Abnormally phosphorylated tau protein
What drugs act on microtubules?	Mebendazole/thiabendazole (antihelminthic), taxol (antibreast cancer), colchicine (antigout), vincristine/vinblastine (anticancer), griseofulvin (antifungal)
What term describes the microtubule complex of flagella and cilia?	Axoneme
What proteins associated with microtubules are involved in neuronal axoplasmic transport?	Kinesins (anterograde) and dyneins (retrograde)

What protein is defective in Kartagener syndrome?	Dynein, which leads to immotile cilia
Describe the structure of a microfilament:	Globular actin monomers (G-proteins) linked to a double helix
Microfilaments are involved with what cellular processes?	Establishing focal contact between the cell and extracellular matrix, locomotion of nonmuscle cells, formation of the contractile ring in dividing cells, folding of epithelia into tubes during development
Microfilaments are an important component of what GI cell structure?	Microvilli
What is the importance of these structures in the GI tract?	Increase the absorptive surface area
Name some key intermediate filaments:	Keratin, desmin, vimentin, neurofilaments, glial fibrillary acidic protein, lamins
What is the role of intermediate filaments?	Providing mechanical strength to cells
What disease, characterized by acantholysis with painful oral ulcers, is a result of autoantibody destruction of either desmoglein or transmembrane *E*-cadherin adhesion molecules?	Pemphigus vulgaris

ORGANELLE INTERACTIONS

What term describes the uptake of materials by a cell?	Endocytosis
Describe the sequence of events occurring in receptor-mediated endocytosis:	Ligand binds specifically to its receptor on the cell surface.
	Ligand-receptor complex clusters into a clathrin-coated pit that invaginates and produces a clathrin-coated vesicle containing the ligand.
	Within the cytoplasm, the clathrin coat is rapidly lost, leaving an uncoated endocytic vesicle containing the ligand.

What genetic defects can cause familial hypercholesterolemia?

Either the inability to synthesize low-density lipoprotein (LDL) receptors or synthesis of defective LDL receptors that cannot bind LDLs and/or clathrin-coated pits

Phagocytosis is commonly a characteristic of what type of cell?

Macrophages

The aforementioned cell binds to bacteria via which of its receptors?

Fc receptors, C3b receptors

Exocytosis requires an interaction between receptors on what two cellular structures?

1. The granule destined for exocytosis
2. The plasma membrane

What are the two forms of exocytosis?

1. Regulated secretion
2. Constitutive secretion

Which form of exocytosis requires an extracellular signal?

Regulated secretion

Describe the concept of membrane recycling:

The secretory granule membrane added to the plasma membrane surface during exocytosis is retrieved during endocytosis via the clathrin-coated vesicles. Thus, the surface area of the plasma membrane remains relatively constant.

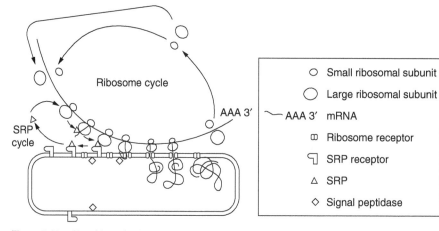

Figure 6.10   Signal hypothesis—explains how proteins that insert into or cross a membrane are synthesized by a membrane-bound ribosome.

Where in the cell are ribosomes, which synthesize membrane-packaged proteins, located?	On the surface of the rough endoplasmic reticulum (RER)
Where are free signal-recognition particles (SRPs) found in the cell?	Cytoplasm
SRPs bind to what portion of the ribosome-mRNA complex?	Signal sequence of mRNA
What structures anchor the polyribosomes-mRNA-SRP complex to the RER membrane following relocation of this complex from the cytosol to the RER?	Ribosome receptor proteins and SRP receptors in the RER membrane
Newly formed polypeptide is threaded through what structure in the RER membrane?	A pore across the RER membrane (thus, the polypeptide is pushed into the RER cisterna)
What happens to the newly synthesized polypeptide once in the RER cisterna?	Signal peptidase cleaves the SRP, polypeptide is glycosylated
Newly synthesized polypeptide is transferred next from the RER cisterna to what organelle?	*cis*-Golgi network (via coatamer-coated vesicles)
In what direction do proteins destined for membrane packing move in the Golgi apparatus?	From the *cis* to the *trans* face of the Golgi apparatus (again via coatamer-coated vesicles)
Do all of the events of protein processing occur in the same compartment of the Golgi apparatus?	No, they occur in distinct cisternal subcompartments
What part of the Golgi apparatus sorts proteins for their final destinations?	*trans*-Golgi network

CELL-TO-CELL COMMUNICATION

Name some examples of signaling molecules:	Neurotransmitters, endocrine hormones, local mediators (paracrine hormones, autocrine hormones), Ca^{2+} ions (intracellularly)
What are the two classes of signaling	1. Lipid-soluble signaling molecules molecules?
	2. Hydrophilic signaling molecules

Where do each of these classes of signaling molecules bind in the cell?	Lipid-soluble molecules penetrate the plasma membrane and bind receptors in the cytoplasm or nucleus
	Hydrophilic molecules bind cell surface receptors
How does each class of signaling molecules create an effect in a cell?	Lipid-soluble molecule-receptor complexes interact directly with DNA-binding sites (therefore creating slow-acting, long-term effects)
	Hydrophilic molecule-receptor complexes are coupled to second messenger systems and phosphorylation of cellular proteins (therefore creating fast-acting, short-term effects)

RECEPTORS

What signaling molecules utilize intracellular receptors?	Steroid molecules like progesterone, estrogen, testosterone, cortisol, aldosterone, as well as thyroxine
Nuclear hormone receptor complexes bind to DNA and stimulate what enzymatic modification of histones?	Histone acetyltransferase acetylates histones resulting in opening/relaxation of DNA, thus facilitating transcription.
	Histone deacetylase deacetylates histones resulting in condensing of DNA, thus preventing transcription.
What amino acid residues compose the DNA-binding regions of steroid hormone receptors?	Cysteine residues
What disorder is characterized by hypertension, truncal obesity, hyperglycemia, buffalo hump, and immune suppression?	Cushing syndrome
What glucocorticoid/progesterone receptor inhibitor has been used to treat the aforementioned disorder, and recently as an abortion method?	Mifepristone (RU486)
What receptor gates the Na^+/K^+ ion channel?	Nicotinic acetylcholine receptor
Antibodies to this receptor results in what disease?	Myasthenia gravis

What drug can be used to treat this disease?	Neostigmine (acetylcholine esterase inhibitor)
Name several signaling molecules that utilize a receptor tyrosine kinase (RTK):	Insulin, epidermal growth factor, platelet-derived growth factor
Describe the chain of events leading to activation of the RTK:	Ligand binds the extracellular domain of the RTK.
	RTK dimerizes with the adjacent RTK.
	Intracellular kinase domains are activated.
	Intracellular autophosphorylation of the RTK.
	Phosphorylated residues become the recognition/anchoring sites for other RTK substrates.
The signal termination of RTK is accomplished via what mechanism?	Internalization of the RTK-signaling molecule complex into an endosome
What disease can be caused by either a decrease in the number of insulin receptors or a dysfunction in the signaling of insulin receptors?	Type 2 diabetes mellitus

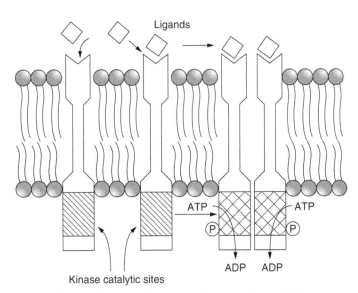

Figure 6.11   Ligand-induced conformation states of RTK.

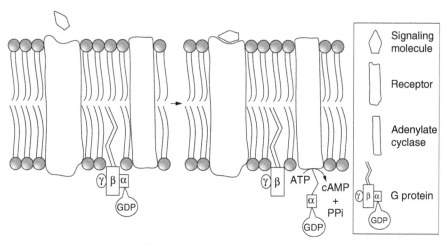

Figure 6.12   Functioning of GPRs.

What class of drugs increase the target cell receptors' response to insulin?	Glitazones (pioglitazone, rosiglitazone, troglitazone)
Name several signaling molecules that utilize G-protein-coupled receptors (GPCRs):	Catecholamines, hormones, local mediators
Describe the chain of events following binding of the signaling molecule to the G_{sp}PCR:	Signaling molecule binds to the receptor. α-Subunit of the G_s protein binds guanosine 5'-triphosphate (GTP), dissociating from the β- and γ-subunits. GTP-α-subunit complex activates adenylate cyclase. cAMP is produced from ATP. cAMP activates PKA (cAMP-dependent PKA).
List the G_sPCRs:	β_1, β_2, D_1, H_2, V_2
What bacteria elaborate toxins that constitutively activate the GPCR?	*Bordatella pertussis* or *Vibrio cholera* via adenosine diphosphate (ADP) ribosylation of the α-subunit
Cellular cAMP levels can be increased by inhibiting what enzyme?	Phosphodiesterase
Name some inhibitors of the aforementioned enzyme:	Caffeine, sildenafil, tadalafil, milrinone, theophylline
What effect do G_iPCRs have on cAMP production?	Decrease cAMP production by inhibiting adenylate cyclase

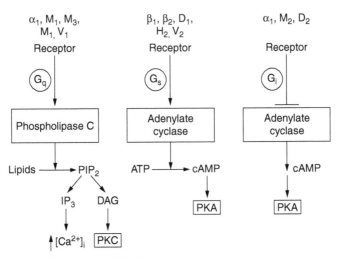

Figure 6.13   GPR-linked second messengers diagram.

List the G_iPCRs:	α_2, M_2, D_2
What enzyme is activated following binding of the signaling molecule to the G_qPCR?	Phospholipase C, leading to formation of inositol triphosphate and diacylglycerol
What are the eventual effectors following G_qPCR activation?	Increased intracellular Ca^{2+}, activation of PKC
List the G_qPCRs:	α_1, M_1, M_3, H_1, V_1

ARACHIDONIC ACID PRODUCTS

What enzyme liberates arachidonic acid from the plasma membrane?	Phospholipase A_2
Name an inhibitor of this enzyme:	Corticosteroids (halobetasol, hydrocortisone, dexamethasone, prednisone)
What toxicity is associated with this inhibitor?	Iatrogenic Cushing syndrome—truncal obesity, moon facies, buffalo hump, muscle wasting, osteoporosis, easy bruisability
What enzyme is responsible for hydroperoxide (ie, leukotriene) production from arachidonic acid?	Lipooxygenase

Name an inhibitor of this enzyme: Zileuton

List the hydroperoxides and their functions: LT B_4 is a neutrophil chemotactic agent; LT C_4, D_4, and E_4 function in bronchoconstriction, vasoconstriction, contraction of smooth muscle, and increasing vascular permeability.

What drugs inhibit hydroperoxide function?   Zafirlukast, montelukast

What enzyme is responsible for endoperoxide (ie, thromboxane, prostaglandin, and prostacyclin) production from arachidonic acid?   Cyclooxygenase (COX-1, COX-2)

Name the inhibitors of this enzyme: Aspirin (irreversibly inhibits COX-1 and COX-2), nonsteroidal anti-inflammatory drugs (NSAIDs) (reversibly inhibit COX-1 and COX-2), celecoxib/rofecoxib (inhibits COX-2 only), acetaminophen (reversibly inhibits COX-1 and COX-2, mostly in the CNS)

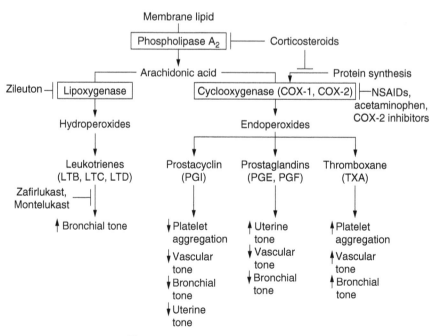

Figure 6.14   Arachidonic acid products.

List the endoperoxides and their functions:	Thromboxane A_2 (TxA_2) stimulates platelet aggregation and vasoconstriction; prostaglandin E (PGE) and prostaglandin F (PGF) increase uterine tone and stimulate vasodilation and bronchodilation; prostaglandin I (PGI) inhibits platelet aggregation, decreases uterine tone, and stimulates vasodilation and bronchodilation.

MUSCLE

What is a neuromuscular junction?	Synapse between a nerve and a muscle cell
What neurotransmitter is typically found in a neuromuscular junction?	Acetylcholine
What toxin prevents exocytosis of secretory vesicles containing this neurotransmitter?	Botulinum toxin (causing flaccid paralysis)
What toxin blocks inhibition of neurotransmitter release?	Tetanus toxin (causing tetanic paralysis)
Name one depolarizing and one nondepolarizing neuromuscular blocking agent:	Succinylcholine (depolarizing), tubocurarine (nondepolarizing)
Name the connective tissue layer which contains the blood supply to skeletal muscle:	Endomysium
What connective tissue layer surrounds individual skeletal muscle fibers?	Perimysium
What connective tissue layer surrounds the bundles of skeletal muscle fibers?	Epimysium
What is a myofiber?	A single multinucleated muscle cell
What are regularly arranged filaments of myofiber termed?	Myofibrils
Describe the boundaries of a sarcomere in the myofiber:	Region between two Z disks
What is the function of the Z disk?	Anchor thin filaments to the sarcomere

What protein composes

Thick filaments?   Myosin

Thin filaments?   Actin

Name the area of the myofiber that   A-band
contains the thick filaments and
stains dark:

Name the area of the myofiber within   H-band
the A-band that stains lighter:

Name the area of the myofiber that   I-band
contains the thin filaments and
stains light:

What is a T-tubule?   The transverse tubule that passes from
                                    the sarcolemma across a myofibril

What is the sarcoplasmic reticulum?   Muscle endoplasmic reticulum
                                    containing Ca^{2+} ion stores

What is a dihydropyridine receptor?   A voltage-gated receptor located on the
                                    plasma membrane of a muscle cell that
                                    acts as Ca^{2+} channel

What is a ryanodine receptor?   A voltage-gated receptor located on the
                                    muscle endoplasmic reticulum that acts
                                    as a Ca^{2+} channel

What disorder is characterized by   Malignant hyperthermia (result of
excessive Ca^{2+} release from the   abnormal opening of the ryanodine
sarcoplasmic reticulum?   receptor)

What drug is used to treat this disorder?   Dantrolene

Describe the sequence of events   Depolarization travels down the T-tubule.
leading to skeletal muscle contraction
following the binding of acetylcholine   The dihydropyridine receptor activates
to the acetylcholine receptor: the ryanodine receptor to release Ca^{2+}.

                                    Released Ca^{2+} binds troponin C.

                                    Troponin C changes conformation.

                                    Tropomyosin moves out of the myosin-
                                    binding groove on the actin filament.

                                    Myosin hydrolyzes its bound ATP and
                                    is displaced on the actin filament
                                    (power stroke).

                                    Contraction causes HIZ shrinkage.

What protein anchors the cortical   Dystrophin
cytoskeleton of a muscle cell to the
transmembrane glycoproteins?

What disease manifests as a result of a defect in the aforementioned protein?	Duchenne muscular dystrophy
Are males or females more likely to develop Duchenne muscular dystrophy?	Only males are afflicted (remember Duchenne muscular dystrophy is an X-linked recessive disorder).
What structures physically connect cardiac muscle cells?	Intercalated discs

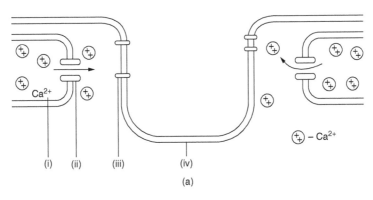

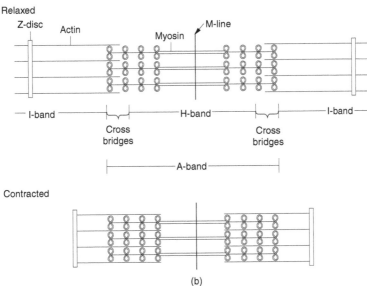

Figure 6.15   Skeletal muscle contraction. (a) Sarcoplasmic reticulum: (i) Sarcoplasmic reticulum, (ii) Ryanodine receptor, (iii) Dihydropyridine receptor, (iv) T-tubule membrane; (b) HIZ shrinkage: (i) Actin thin filament, (ii) Myosin thick filament.

What structures electrically couple cardiac muscle cells?

Gap junctions

What is meant by the term *calcium-triggered calcium release* in regards to cardiac muscle cell contraction?

Contraction is dependent on extracellular Ca^{2+}, which enters the cells during the plateau of an action potential and stimulates the release of sequestered Ca^{2+} from the sarcoplasmic reticulum.

What two proteins remove excess Ca^{2+} from the cytoplasm of a cardiac muscle cell following a contraction cycle?

1. Ca^{2+}-ATPase
2. Na^+/Ca^{2+} exchanger

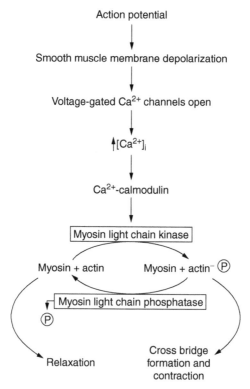

Figure 6.16 Smooth muscle contraction.

What drug is used to increase cardiac contractility indirectly by inhibiting the removal of excess Ca^{2+}?	Digitalis (ouabain)
What ion-protein complex activates myosin light chain kinase in smooth muscle cells?	Ca^{2+}-calmodulin
Does myosin light chain kinase activation result in contraction or relaxation?	Contraction
Increased intracellular levels of what second messenger will typically result in smooth muscle relaxation?	Cyclic guanosine monophosphate (cGMP)
What drug, used to treat male impotence, reduces breakdown of the aforementioned second messenger?	Viagra (cGMP phosphodiesterase inhibitor)
What drug, used to treat severe hypertension, reduces afterload?	Hydralazine (vasodilates arterioles > veins)

SUMMARY CHART

Chart 6.1 Lysosomal Storage Diseases

Disease	Inheritance	Deficient Enzyme	Substrate Accumulation	Signs/Symptoms
Fabry	X-linked recessive	α-Galactosidase A	Ceramide trihexoside (glycosphingolipid)	- Anhidrosis (lack of sweating) - Fatigue - Angiokeratomas - Burning extremity pain - Ocular involvement
Krabbe (globoid leukodystrophy)	Autosomal recessive	β-Galactosidase	Galactocerebroside	- Multinucleated globoid macrophage cells
Gaucher disease Type I	Autosomal recessive	β-Glucocerebrosidase	Glucocerebroside	- Massive splenomegaly -Erlenmeyer flask bone lesions & pathologic fractures - Pancytopenia - Thrombocytopenia - PAS-positive cellular inclusions
Gaucher disease Type II	Autosomal recessive	β-Glucocerebrosidase	Glucocerebroside	- Acute neuronopathy - Hepatosplenomegaly - CNS involvement - Convulsions - Mental deterioration - Death at a young age

Disease	Inheritance	Accumulated Substance	Clinical Features
Niemann-Pick	Autosomal recessive	Sphingomyelin	- Delayed development - Decreased visual acuity - Neuronal death and shrinkage of the brain - Massive hepatosplenomegaly - Infiltration of bone marrow - Death by 3 years of age - Cherry-red spot on the macula
Tay-Sachs	Autosomal recessive	G_{M2} ganglioside	- Zebra bodies (seen histologically) - Motor and mental retardation - Blindness - Hyperacute hearing - Death usually by 2–3 years of age - Cherry-red spot on the macula
Metachromatic leukodystrophy	Autosomal recessive	Sulfatide in the brain, kidney, liver, and peripheral nerves	- Spasticity with demyelination and gliosis - Increased CSF protein - Ataxia - Dementia

(Continued)

Chart 6.1 Lysosomal Storage Diseases (*Continued*)

Disease	Inheritance	Deficient Enzyme	Substrate Accumulation	Signs/Symptoms
Hurler	Autosomal recessive	α-L-Iduronidase	Heparan and dermatan sulfate	- Coarse facial features - Hepatosplenomegaly - Lesions of cardiac valves - Narrowing of the coronary arteries - Joint stiffness - Kyphoscoliosis - Mental retardation - Corneal clouding
Hunter	X-linked recessive	Iduronate sulfatase	Heparan and dermatan sulfate	- Milder than Hurler - Aggressive behavior - No corneal clouding

CLINICAL VIGNETTES—MAKE THE DIAGNOSIS

A 9-month-old is seen by her pediatrician and is noted to have psychomotor retardation along with a palpable liver and spleen. Ophthalmologic examination revealed a cherry-red spot in the macula. Pulmonary and cardiac examinations were unremarkable. Bone marrow examination revealed the presence of numerous foam cells.

This patient has what disease?	Niemann-Pick disease
This disorder particularly affects which ethnic group?	Ashkenazi Jews
Death usually results from what?	Neurological damage

A 6-month-old baby boy of Ashkenazi Jewish decent is evaluated by a pediatrician because of loss of motor skills and decreased attentiveness. Upon physical examination, the baby is found to have hyperreflexia with sustained ankle clonus. His liver and spleen were not palpable. Ophthalmologic examination revealed a cherry-red spot in the macula. Serum enzyme analysis reveals the absence of detectable activity in hexosaminidase A.

This enzyme degrades what substance?	Ganglioside G_{M2}
Intracellular accumulations occur in which cytoplasmic organelle?	Lysosome
Histological examination would reveal what characteristic finding in nerve cells?	Fat droplets

A thin-appearing 3-year-old white girl is brought by her parents to the pediatrician's office because of constant difficulty in breathing. She has a consistent cough that is productive of green sputum and causes difficulty in sleeping. The parents note that the child has had foul-smelling stool since birth and frequent urinary tract infections.

The child likely is showing signs of what underlying disease?	Cystic fibrosis
How is this disease inherited?	Autosomal recessive
What test is routinely used to screen for this disease?	Chloride sweat test

A 4-year-old white male is brought by his parents to the pediatric clinic because of easy fatigability and difficulty in walking of several months' duration. Concurrently, the child's calves have increased in size. The mother reports that the child has begun to "climb on himself" to rise from a sitting position.

This child likely has what disease?	Duchenne muscular dystrophy
The child inherited his defective gene from which parent?	The mother (X-linked recessive disorder)
The disease is characterized by marked deficiency of what protein?	Dystrophin (stabilizes actin filaments in the muscle)

A 32-year-old male complains of diminishing vision of his left eye. Additionally, he reports painful, burning sensations of his palms and soles. His family history is significant for chronic renal failure. Physical examination reveals angiokeratomas present on the skin around his umbilicus, buttocks, and scrotum. There is a corneal leukomatous opacity and retinal edema in the left eye. There is edema in the lower extremities.

What disease is this patient suffering from?	Fabry disease
What is the etiology behind this disease?	X-linked recessive disorder of glycosphingolipid metabolism caused by a deficiency of α-galactosidase A
What is a potentially fatal complication of this disease?	Renal failure

A 21-year-old mentally retarded Jewish male presents to his primary care provider with complaints of recurrent epistaxis and easy bruising. Directed questioning reveals weakness and an enlarging left-sided abdominal mass. Physical examination reveals multiple purpuric patches, pallor, mild hepatomegaly, and massive splenomegaly. Bone marrow biopsy shows "wrinkled paper" intracytoplasmic inclusions that stain *p*-aminosalicylic acid (PAS) positive.

This patient has been suffering from what disorder?	Gaucher disease
What is the deficient enzyme and mode of inheritance of this disorder?	Glucocerebrosidase, autosomal recessive
What do the "wrinkled paper" inclusions represent?	Glucocerebroside deposition

A 9-year-old male with a course face and large tongue is brought into the clinic because of decreased hearing. Physical examination reveals no corneal opacities. Urine analysis shows increased heparin sulfate and dermatan sulfate.

From what disease is this child suffering and what is its etiology?	X-linked recessive Hunter disease (type II mucopolysaccharidosis), iduronosulfate sulfatase deficiency
Leukocytes will likely show what distinguishing characteristic with microscopic analysis?	Reilly bodies (metachromatic granules)
What auditory complication is associated with this disorder?	Deafness

A 20-month-old mentally and physically delayed male presents with bilateral corneal clouding. The child showcases *gargoylism* with coarse, elongated facial features. Physical examination reveals cardiomegaly, hepatomegaly, and kyphoscoliosis.

This child suffers from a deficiency of what enzyme?	α-Iduronidase (Hurler syndrome, type I mucopolysaccharidosis)
What is the mode of inheritance of this disorder?	Autosomal recessive
What abnormal substances would be found in this patient's urine?	Chondroitin sulfate B, heparin sulfate

A 6-month-old infant is brought to the clinic because of failure to thrive. The mother reports a weaker suck and increased regurgitation as compared to her previous children. Physical examination is notable for underdevelopment, global spasticity, and hyperactive deep tendon reflexes. Microscopy reveals basophilic multinucleated macrophages with cytoplasmic inclusions.

This child is suffering from what disorder?	Krabbe disease (globoid leukodystrophy)
What is the deficient enzyme and mode of inheritance of this disorder?	Galactosylceramide β-galactosidase, autosomal recessive
What are the typical central nervous system (CNS) findings?	Demyelination of cerebral, cerebellar, and basal ganglier white matter

A 4-year-old white male is brought to the pediatrician by his parents because of generalized "stiffness" and difficulty walking, especially climbing stairs, for the past several months. Physical examination reveals a child with a wide-based unsteady gait accompanied by ataxia. Deep tendon reflexes are exaggerated and Babinski sign is elicited. Microscopy reveals brownish granules in oligodendrocytes upon staining with toluidine blue.

This child is suffering from what disease process?	Metachromatic leukodystrophy (autosomal recessive)
What is this patient's enzyme deficiency?	Arylsulfatase A
In what additional organ will this patient show the aforementioned deposition?	Kidneys

A 22-year-old male presents with complaints of recurrent sinus and upper respiratory tract infections. They have always been a problem, but these infections have caused him to miss a substantial amount of work at his new job. On physical examination, it is noted that his heart is on the right side of his chest. Additionally, his liver is on the left side of his abdomen.

This patient can be diagnosed with what condition?	Kartagener syndrome
Why has this patient experienced recurrent sinus and upper respiratory tract infections?	The lack of dynein renders cilia of the sinuses and bronchi immotile, thus there is no mucus-clearing action.
Why would this patient also be infertile?	His spermatozoa would be immotile as a result of the lack of dynein.

A 1-year-old white girl is brought in by her parents because of lethargy, weakness, and yellowing of her skin. The child has palpable splenomegaly. CBC reveals microcytic anemia with dark red blood cells (RBCs) lacking any central pallor and increased numbers of reticulocytes. The father reports that he had been diagnosed with a blood disorder as a child.

This child is suffering from what disease?	Hereditary spherocytosis
Why do the RBCs in this disease have an abnormal, spherical shape?	The RBC membranes lack spectrin, necessary to maintain the normal biconcave disk shape.
What is the significance of the splenomegaly?	The abnormal shape of RBCs leads to higher rates of splenic sequestration and hemolysis with elevated direct bilirubin.

A woman who survived a fire with traumatic full-body burns is resting in the hospital in stable condition. Suddenly, the patient begins to use her accessory muscles of respiration and experiences dyspnea, tachypnea, rales, wheezing, and rhonchi bilaterally. CXR shows diffuse alveolar densities and air bronchograms.

This patient is experiencing what syndrome?	Acute respiratory distress syndrome
What is the etiology behind this patient's syndrome?	Decreased surfactant production, neutrophil release of proteases, free radicals, and leukotrienes; lymphocyte and macrophage release of interleukin-1 and tumor necrosis factor
What is the immediate treatment of this disorder?	Mechanical ventilation, intensive care unit (ICU) care

Index

Note: The numbers followed by "*f*" and "*t*" denote figures and tables, respectively.